# Applied Methods in Design and Construction of Bridges, Highways and Roads - Theory and Practice

*Edited by Khaled Ghaedi*

Published in London, United Kingdom

Applied Methods in Design and Construction of Bridges, Highways and Roads - Theory and Practice
http://dx.doi.org/10.5772/intechopen.95678
Edited by Khaled Ghaedi

Contributors
Abdul Ridha Mohammed Afrawee, Hussein Ali Mohammed, Haider Habeeb Aodah, Nilton de Souza Campelo, Arlene Maria Lamêgo da Silva Campos, Marcos Valério Mendonça Baia, Daniel Jardim Almeida, Raimundo Humberto Cavalcante Lima, Danielly Kelly dos Reis Dias, Júlio Augusto de Alencar Júnior, Mário Jorge Gonçalves Santoro Filho, David Garcia-Sanchez, Alvaro Gaute-Alonso, Eva Burgetová, Denisa Boháčová, Meisam Gordan, Saeed-Reza Sabbagh-Yazdi, Khaled Ghaedi, Zubaidah Ismail, David P. Thambiratnam, Owuama C. Ozioma, Júlio Alencar Jr, Ahad Javanmardi, Hamed Khatibi, Ramin Veghei



First published in London, United Kingdom, 2022 by IntechOpen
IntechOpen is the global imprint of INTECHOPEN LIMITED, registered in England and Wales, registration number: 11086078, 5 Princes Gate Court, London, SW7 2QJ, United Kingdom

British Library Cataloguing-in-Publication Data
A catalogue record for this book is available from the British Library

Additional hard and PDF copies can be obtained from orders@intechopen.com

Applied Methods in Design and Construction of Bridges, Highways and Roads - Theory and Practice
Edited by Khaled Ghaedi
p. cm.
Print ISBN 978-1-80355-555-3
Online ISBN 978-1-80355-556-0
eBook (PDF) ISBN 978-1-80355-557-7

# Meet the editor

Dr. Ghaedi has experience in both academia and industry. He has published articles in prestigious journals that have produced novel patented damping devices, guidelines for structural retrofitting, and versatile models to predict structural damages. He has contributed to the process of knowledge sharing through his roles as editor, reviewer, speaker, and committee member for journals, conferences, and different organizations. Dr. Ghaedi is also a committee member of the well-known international design code ASCE 7-22. He is currently supporting the engineering industry in Southeast Asia as a structural technical director. He is the founder of the PASOFAL Engineering group, a talented team recognized for solving complex engineering challenges. He is highly skilled in research funding acquisition, software training, and design consultation through Finite Element Methods, Computational Fluid Dynamics, Structural Health Monitoring, Building Information Modeling, Performance-Based Design, and other organizations.

# Contents

# Preface

Civil structures and infrastructure, particularly roadways and bridges, face multiple lifelong obstacles caused by environmental impacts, such as corrosion, microstructural defects, cracks, thermal and residual stresses, instability, bond failures, and natural disasters like earthquakes and floods. With the advancement of the transportation industry, the scale of bridge, highway, and road construction is growing and construction quality has become an essential factor in transportation projects. The quality of bridge, highway, and road works is not only influenced by material choice, construction team, and management quality but also technology. This book presents new solutions in bridge, highway, and road design and construction to improve the quality of bridge, highway, and road safety.

The book begins with a brief introduction to the design, construction, and retrofitting of bridges, highways, and roads. For the design and construction of bridge structures, it presents current optimization methods using soft computing techniques to address sustainable design and construction of bridges. In addition, the book discusses gaps in bridge optimizations and proposes suggestions for future research. It also introduces an innovative approach for assessing the cross-sectional distribution of live loads on bridge beam-deck systems, reducing the structural problem from multiple degrees of freedom system to two degrees of freedom system. The book investigates the reconstruction of historic bridge deck systems using a case study as well as highway construction challenges. It discusses highway management systems in Iraq that require serious examination and modification. Finally, the book proposes a sustainable trenchless drainage system for road construction on flat terrains.

I hope that readers will find this book useful and I warmly welcome comments, suggestions, and criticisms.

**Khaled Ghaedi, Ph.D.**
PASOFAL Engineering Group,
Kuala Lumpur, Malaysia

Chapter 1

# Introductory Chapter: Design, Construction, and Retrofit of Bridges, Roads, and Highways

*Khaled Ghaedi, Meisam Gordan, Ahad Javanmardi, Hamed Khatibi and Ramin Vaghei*

## 1. Introduction

Bridges are structures designed to support traffic and other dynamic loads induced by vehicle loads to pass through natural or manmade obstacles. Types of pathways may be roads, highways, railroads, pipeline waterways, or pedestrians. Obstacles can be categorized as canals, rivers, mountains, valleys, lakes, seas, and other manmade structures, such as buildings, rail lines, roads, and bridges themselves. A bridge is a vital structure of modern roadway and railway systems and largely serves as the lifeline of public infrastructure. Apart from bridge structures, Roads and highways are commonly considered important to new-fashioned life and they play an important role in the advancement of cities. The lack of quantifiable sustainability methods creates gaps in sustainability knowledge, leading to the public, and environmental and financial dissatisfaction with completed highways and urban roads. This chapter discusses the key points in design and construction assessment of bridges, highways, and roads covered in the present book in order to bridge the knowledge gap. The discussed topics such as bridge optimization techniques, risk assessment of roads, highway management systems, and challenges in highway construction presented in the book chapters help to provide insights into the bridges and roads' impacts on the environment and the benefits of adopting development assessment systems to increase safety of bridge structures, roads, and highways.

### 1.1 Challenges and solutions

Civil structures and infrastructure, in particular roadways and bridges, have multiple lifelong obstacles caused by various reasons of environmental impact such as corrosion, microstructural defects, cracks, thermal and residual stresses, instability, bond failures, or natural disasters such as earthquake and flood [1–3]. Damage to structural elements affects structural properties such as mass, stiffness, and damping, resulting in changes in the dynamic response of the structure such as natural frequency, modal shape, and damping ratio [4–9]. Therefore, actual solutions can play an essential role in ensuring the safety and reliability of structures. In recent decades, the advancement of road and bridge structures along with their optimization in design and construction has attracted much attention. Development and structural optimization based on mathematical and numerical analysis have resulted in strategies

 IntechOpen

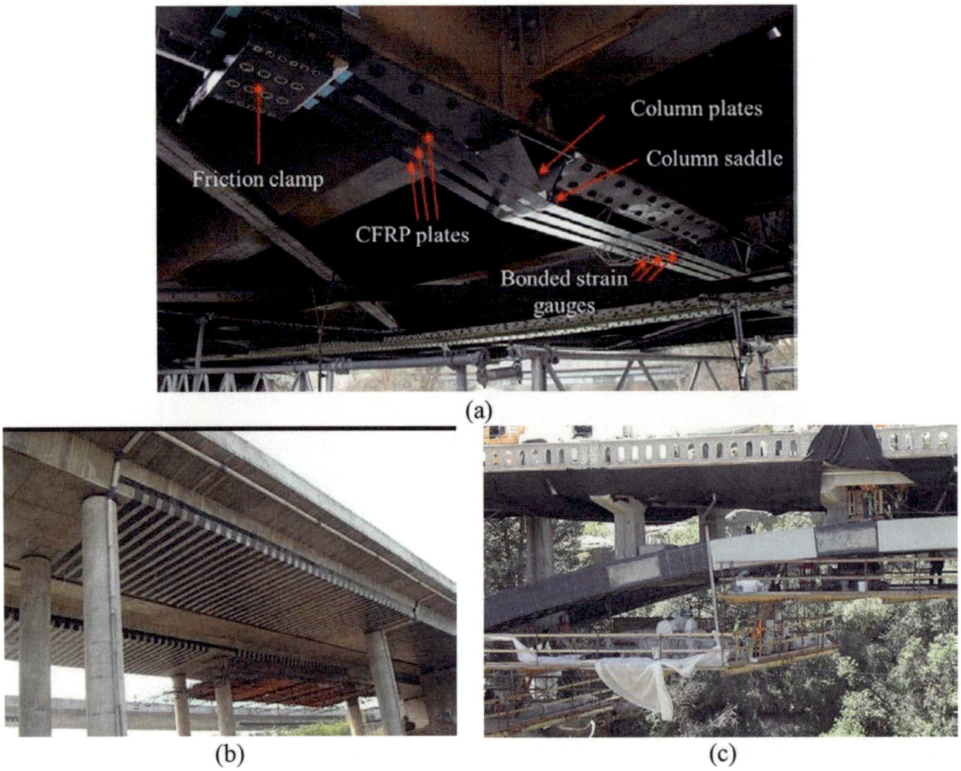


**Figure 1.**
*Bridge strengthening with (a) CFRP plates, (b) CFRP strips and (c) CFRP sheets [19].*

applied primarily for fruitful and sustainable design in road and bridge construction [10–17]. For instance, the strength of bridge elements can be significantly increased using Carbon fiber-reinforced polymer (CFRP) composites due to their outstanding performance [18]. **Figure 1** shows a few examples of bridge strengthening using CFRP. As mentioned, new tools and technologies can also be adapted to design new bridges and assess and retrofit existing bridge structures [20, 21]. **Figure 2** shows the utilization of advanced tools and techniques in damage assessment of a slab-on-girder bridge structure utilizing the structural health monitoring (SHM) approach.

Roads, as another important traveling pass, provide a means to communicate between cities or even countries. They can be designed as a one-lane road or a multiple-lane road that aligns or intersects with each other. During construction, different requirements have to be made to provide an effective solution. One of those requirements is road drainage systems in order to maintain the structural integrity of the roadway [22]. This measure guarantees road usability and sustainability over the long term. Drainage channels on one or both sides of the road provide a means of harmlessly diverting outflows from road surfaces and neighboring facilities to artificial or natural drainage channels [23]. **Figure 3** depicts examples of drainage systems for roads and highways.

Inappropriate slope stability can lead to road drainage problems and cause road damage. Material flowing to the channel bottom may block the flow of water into the channel and cause water to leak into the road body. This causes discrepancies and shoulder deformity. **Figure 4** demonstrates how an unstable raceway slope fills the

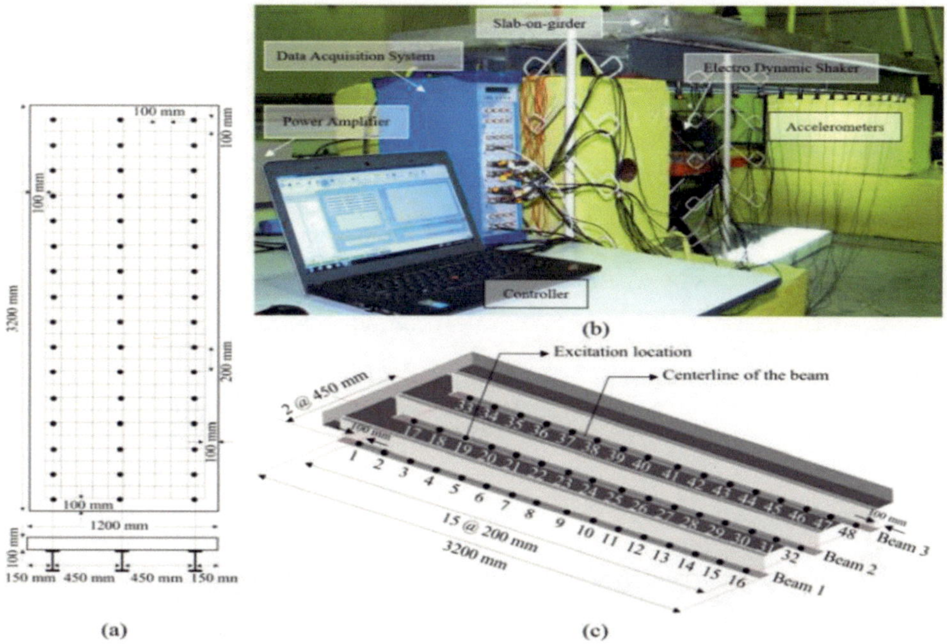


**Figure 2.**
*SHM technique to detect bridge damages using vibration test method as (a) schematic view and dimensions of the slab-on-girder bridge, (b) experimental setup of the test and (c) the arrangement of the accelerometers and the excitation point.*

**Figure 3.**
*Drainage systems.*

bottom of the channel and causes an annual increment in road surface settlement. Furthermore, **Figure 5** shows unlined channels that cause severe destruction to the channel in the form of silting and washout.

Despite extensive data, recent developments in structural design, construction, and retrofitting have not yet been fully explored. Therefore, the main purpose of this book is to review past studies on bridges, highways, and roads to provide a detailed investigation in order to come up with reasonable solutions for existing challenges in design and construction using advanced methods and technologies. It also helps engineers better understand designing new bridges, roads, and highways, and retrofitting techniques for previously constructed of such systems. This book also proposes a sustainable drainage system for road construction.

**Figure 4.**
*Unstable raceway slope fills the bottom of the channel and causes an annual increment in road surface settlement.*

**Figure 5.**
*Road channel fracture due to lack of proper slope angle.*

## Author details

Khaled Ghaedi[1*], Meisam Gordan[1,2], Ahad Javanmardi[1,3], Hamed Khatibi[1,4,5] and Ramin Vaghei[1,6]

1 Research and Development Center, PASOFAL Engineering Group, Kuala Lumpur, Malaysia

2 Department of Civil Engineering, Universiti Putra Malaysia (UPM), Selangor, Malaysia

3 College of Civil Engineering, Fuzhou University, Key Lab of Fujian Province, Fuzhou, China

4 The University of Auckland, New Zealand

5 Aimecs Engineering Ltd, New Zealand

6 Civil Engineering Department, Islamic Azad University of Mashhad, Mashhad, Iran

*Address all correspondence to: khaledqhaedi@yahoo.com

## References

[1] Hanif MU, Ibrahim Z, Ghaedi K, Hashim H, Javanmardi A. Damage assessment of reinforced concrete structures using a model-based nonlinear approach—A comprehensive review. Construction and Building Materials. 2018;**192**:846-865

[2] Ghaedi K, Gordan M, Ismail Z, Hashim H, Talebkhah M. A literature review on the development of remote sensing in damage detection of civil structures. Journal of Engineering Research Reports. 2021;**20**(10):39-56

[3] Gordan M, Chao OZ, Sabbagh-Yazdi S-R, Wee LK, Ghaedi K, Ismail Z. From cognitive bias toward advanced computational intelligence for smart infrastructure monitoring. Frontiers in Psychology. 2022;**13**:846610

[4] Gordan M et al. Data mining-based damage identification of a slab-on-girder bridge using inverse analysis. Measurement. 2020;**151**:107175

[5] Javanmardi A, Ibrahim Z, Ghaedi K, Khan NB, Benisi Ghadim H. Seismic isolation retrofitting solution for an existing steel cable-stayed bridge. PLoS One. 2018;**13**(7):e0200482

[6] Ghaedi K, Ibrahim Z, Jameel M, Javanmardi A, Khatibi H. Seismic response analysis of fully base-isolated adjacent buildings with segregated foundations. Advanced Civil Engineering. 2018;**2018**:1-21

[7] Ghaedi K, Ibrahim Z, Javanmardi A, Rupakhety R. Experimental study of a new bar damper device for vibration control of structures subjected to earthquake loads. Journal of Earthquake Engineering. 2018;**25**(2):300-318

[8] Ghaedi K, Ibrahim Z, Javanmardi A. A new metallic bar damper device for seismic energy dissipation of civil structures. Materials Science and Engineering. 2018;**431**:122009

[9] Javanmardi A, Ghaedi K, Huang F, Hanif MU, Tabrizikahou A. Application of structural control systems for the cables of cable-stayed bridges: State-of-the-art and state-of-the-practice. Archives of Computational Methods in Engineering. 2022;**29**(3):1611-1641

[10] Gordan M et al. State-of-the-art review on advancements of data mining in structural health monitoring. Measurement. 2022;**193**:110939

[11] Li T, Lai X, Xiang J, Sun H, Lei D, Xu S. Ecological impacts of sea-crossing bridge construction on local sediment microbiome in East China. Regional Studies in Marine Science. 2022;**53**:102363

[12] Kagioglou P, Katakalos K, Mitoulis SA. Resilient connection for accelerated bridge constructions. Structure. 2021;**33**:3025-3039

[13] Mehdi Kashani M, Saiidi S, Eberhard MO. Resilience-based design for next-generation bridge design and construction. Structure. 2021;**34**:4466

[14] Zaheer Q, Yonggang T, Qamar F. Literature review of bridge structure's optimization and it's development over time. International Journal for Simulation and Multidisciplinary Design Optimization. 2022;**13**:5

[15] Javanmardi A, Ibrahim Z, Ghaedi K, Jameel M, Hanif MU, Gordan M. Seismic response of a base isolated cable-stayed bridge under near-fault ground motion excitations. Scientific Research Journal. 2018;**15**(1):1-14

[16] Ghaedi K, Ibrahim Z, Adeli H, Javanmardi A. Invited review: Recent developments in vibration control of building and bridge structures. Journal of Vibroengineering. 2017;**19**(5):3564-3580

[17] Gordan M, Razak HA, Ismail Z, Ghaedi K. Data mining based damage identification using imperialist competitive algorithm and artificial neural network. Latin American Journal of Solids Structures. 2018;**15**(8):e107

[18] Ghaedi K et al. Finite element analysis of a strengthened beam deliberating elastically isotropic and orthotropic cfrp material. Journal of Civil Engineering Science and Technology. 2018;**9**(2):5

[19] Sonnenschein R, Gajdosova K, Holly I. FRP composites and their using in the construction of bridges. Procedia Engineering. 2016;**161**:477-482

[20] Gordan M, Ghaedi K, Ismail Z, Benisi H, Hashim H, Ghayeb HH. From conventional to sustainable SHM: implementation of artificial intelligence in the Department of Civil Engineering, University of Malaya. In: 2021 IEEE International Conference on Artificial Intelligence in Engineering and Technology (IICAIET). pp. 1-6

[21] Gordan M, Ismail ZB, Abdul Razak H, Ghaedi K, Ghayeb HH. "Optimization-based evolutionary data mining techniques for structural health monitoring." Journal of Civil Engineering Construction. Feb 2020;**9**(1):14-23

[22] Ojha RP, Verma CL, Denis DM, Singh CS, Kumar M. Modification of inverse auger hole method for saturated hydraulic conductivity measurement. Journal of Soil and Water Conservation. 2017;**16**(1):47

[23] Owuama CO. Sustainable drainage system for road networking. International Journal of Innovative Management and Technology. 2014;**5**(2):83

Chapter 2

# Introduction to Monitoring of Bridge Infrastructure Using Soft Computing Techniques

*Meisam Gordan, Saeed-Reza Sabbagh-Yazdi, Khaled Ghaedi, David P. Thambiratnam and Zubaidah Ismail*

## Abstract

More than a billion structures exist on our planet comprising a million bridges. A number of these infrastructures are near to or have already exceeded their design life and maintaining their health condition is an engineering optimization problem. Besides, these assets are damage-prone during their service life. This is due to the fact that different external loads induced by the environmental effects, overloading, blast loads, wind excitations, floods, earthquakes, and other natural disasters can disturb the serviceability and integrity of these structures. To overcome such bottlenecks, structural health monitoring (SHM) systems have been used to guarantee the safe functioning of structures to make satisfactory decisions on structural maintenance, repair, and rehabilitation. However, conventional SHM approaches such as virtual inspections cannot be used for structural continuous monitoring, real-time and online assessment. Therefore, soft computing techniques can be significantly used to mitigate the aforesaid concerns by handling the qualitative analysis of the complex real world behavior. This chapter aims to introduce the optimized SHM-based soft computing techniques of bridge structures through artificial intelligence and machine learning algorithms in order to illustrate the performance of advanced bridge monitoring approaches, which are required to maintain the health condition of infrastructures as well as to protect human lives.

**Keywords:** structural health monitoring, bridge monitoring, optimization, damage assessment, bridge failure, soft computing, artificial intelligence, evolutionary algorithms

## 1. Introduction

Infrastructure monitoring is one of the most significant applications of cities [1]. This is likely due to the fact that smooth functioning of cities is a vital need by providing safe and efficient infrastructure. Eventually, complex, large, and expensive engineering assets, i.e., high-rise buildings, long-span bridges, dams, oil platforms, wind turbines, offshore structures, roadways, and rail tracks are designed to last long [2, 3]. For example, bridges are normally built to have a lifespan of 50 years

[4]. However, many of them are near to or have already exceeded their design life. For example, according to ASCE 2021 infrastructure report card, 42% of all bridges across the United States are at least 50 years old, and they are considered structurally deficient; they are in poor condition and in need of repair. Besides, these assets are damage-prone during their service life. This is due to the fact that different external loads induced by environmental effects, overloading, blast loads, wind excitations, floods, earthquakes, and other natural disasters can disturb the serviceability and integrity of these structures.

Structural and materials design is a highly iterative process for the optimal design of infrastructures. Even the simplest structures and materials are composed of multiple elementary structural components which can lead to various optimal designs [5]. Likewise, maintaining the health condition of infrastructures is also an engineering optimization problem. This is because it is not easy to find an exact solution [6]. For instance, evolutionary techniques have been applied as a part of the procedure of achieving the exact solution. Therefore, several metaheuristic algorithms such as genetic algorithm, particle swarm optimization, ant colony optimization, and imperial competitive algorithm have been developed to solve a variety of engineering optimization problems in a transdisciplinary field of engineering, so-called structural health monitoring (SHM). Bridge monitoring and optimization are significant areas of SHM and soft computing, respectively.

SHM and soft computing techniques as powerful tools can be significantly used to mitigate the aforesaid concerns by planning scheduled maintenance, control, and management of infrastructures. Based on the above explanations, this chapter aims to introduce the optimized SHM-based soft computing techniques of bridge structures through artificial intelligence and machine learning algorithms in order to illustrate the performance of advanced bridge monitoring approaches which are required to maintain the health condition of infrastructures as well as to protect human lives.

## 2. Bridge as an iconic infrastructure

Civil engineering is an ancient profession and one of the most noble in the world. Many structures are monuments to civilization and last for centuries, becoming pilgrim and tourist attractions. Many of these huge monumental structures, being one-off in nature, have warranted large realization times. Some of them have taken centuries to build. The above structural attributes, being passed down from generation to generation, necessarily had certain design processes, motivations, and contexts for posterity to appreciate and admire in terms of form, esthetics, and sustenance [7]. Peter Rice [8], one of the most original and influential engineers of the twentieth century was certain that if the method of manufacturing was true to the project's nature, it would result in a structure capable of producing the emotional response intended by the designer. In the editorial headline "Winning the Emotional Argument," New Civil Engineer editor Mark Hansford commented in 2017: "civil engineering professionals now, more than ever, need to engagingly present the broader benefits of their infrastructure projects, highlighting the direct impacts they are having on society" [9].

Aside from the implicit recognition that successful infrastructure projects now employ a vast range of complementary disciplines—including engineering, planning, architecture, landscape design, ecology, and many others, and are no longer simply

**Figure 1.**
*Evolution of bridge design: Influence of materials on esthetics and structural design. Roman aqueduct in Spain, 50 A.D. [5]. Ponte Vecchio bridge in Italy, 1345 [13]. Büyükçekmece bridge in Turkey, 1567 [14]. Khaju bridge in Iran, 1650 [15]. Pulteney bridge in UK, 1769 [12]. Maria Pia bridge in Portugal, 1877 [5]. London Tower bridge in UK, 1894 [14]. Chengyang bridge in China, 1912 [16]. Salginatobel bridge in Switzerland, 1930 [17]. Sydney Harbour bridge in Australia, 1932 [14]. Golden Gate bridge in USA, 1937 [5]. Coronado Girder bridge in USA, 1967 [5]. Magdeburg Water bridge in Germany, 2003 [18]. Seri Wawasan bridge in Malaysia, 2003 [12]. Helix bridge in Singapore, 2004 [14]. Millau Viaduct bridge in France, 2004 [19]. Shenyang Sanhao bridge in China, 2008 [7]. Rewa Rewa bridge in New Zealand, 2010 [20] Lower Hatea Crossing in New Zealand, 2011 [10]. Rzeszów circular footbridge in Poland, 2012 [21].*

"heavy engineering," it seems self-evident that the bigger the project, the greater the need to demonstrate the benefits to society. It is understood that value must be added—beyond the base metrics of people movement—in terms of lasting social, economic, and environmental benefits [10].

A report by Beade-Pereda [11] stated that bridges link previously separate geographic areas, defying gravity and transforming the landscape. They bind communities, acting as connectors of people, inviting interaction and integration. Bridges are a paradigmatic case of human transformation of nature, symbols of union, progress, and often innovation. Bridges across physical, cultural, and spiritual barriers, frequently becoming landmarks or even icons. They are much more than a piece of infrastructure and the design of such emotional, prominent, and long-lasting constructions should go beyond their main function as structures that link areas, and should always aspire to improve the quality of the built world. At International Association for Bridge and Structural Engineering (IABSE) "Future of Design" event in London on September 8, 2016, Professor Enzo Siviero, named "The Bridge-man" said: "I always say if a bridge was a woman I could marry it!". This hilarious example shows the connection between bridges, emotions, culture, and people [12].

Structural and material design is a highly iterative process for the optimal design of infrastructures. Even the simplest structures and materials are composed of multiple elementary structural components, which can lead to various optimal designs [5]. The evolution in bridge design offers an example of this influence between structural components, materials, and boundary conditions on the design (see **Figure 1**). As can be observed from this figure, the designers have integrated esthetics into the design. Another side of this evolution is dramatic changes in the lighting of bridges in the last decades, which have gained insight into the visual and emotional effects [22].

## 3. Bridge failure statistics: Where? Why? When?

Collapse of bridge structures is one of the worst infrastructural disasters. For example, **Figure 2** shows several spectacular bridge catastrophic collapse cases in the world. It has been reported that the most bridge collapses occurred in North America and Europe, especially in the USA (see **Figure 3**). Extreme conditions under natural hazards, design error and wrong assumption in load combination or ground condition, overloading, impact of vehicles, poor workmanship and inadequate maintenance action, vandalism (e.g., fire and explosion), deterioration, corrosion and fatigue in materials, and limited knowledge due to unknown phenomena are possible causes of bridge collapse. For the sake of clarity, **Figure 4** shows a summarization of the causes of the bridge failure in China. Bridge failure can be occurred both during the service life of the infrastructure or in the construction stage. For example, **Figure 5** tries to illustrate the number of bridge collapses and casualties between 2009 and 2018 [31]. It should be noted that it was difficult to find out more about the prevalence of bridge collapse in Africa, Asia, and South America. **Figure 6** lists a variety of factors, which are essentially not considered in the calculation of the probability of failure.

For better understanding, **Figure 7** presents the catastrophic bridge failures from 1967 until now in the U.S. along with the cause of failures (e.g., corrosion, scour, human error (i.e., design error and construction error), fire, etc.), number of injuries and fatalities.

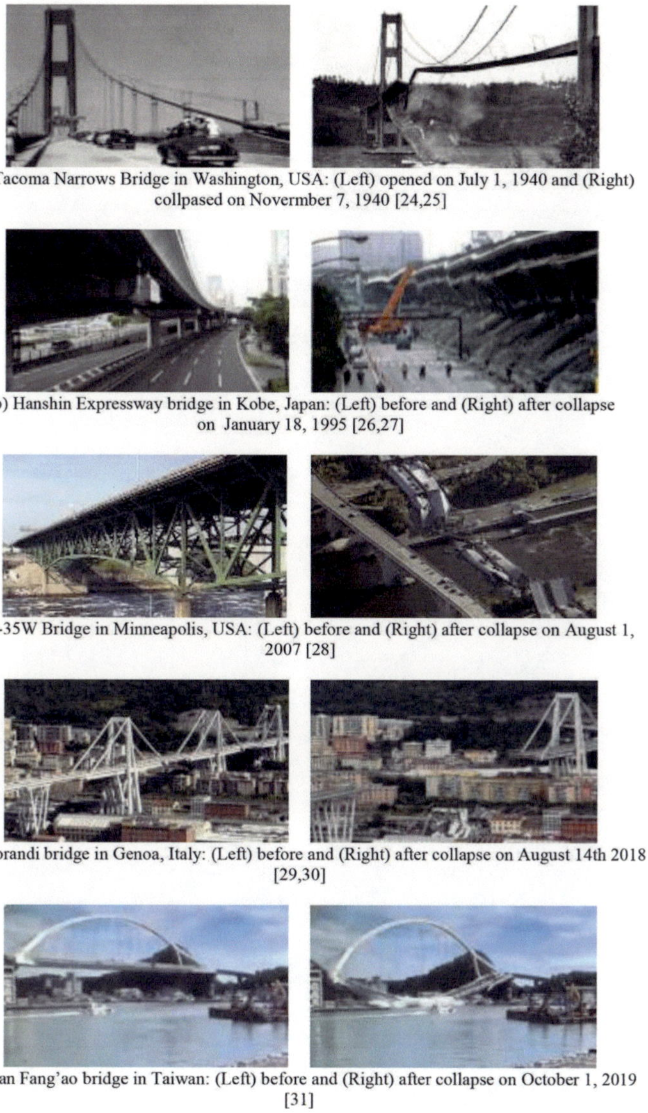


**Figure 2.**
*Examples of bridge catastrophic failure. (a) Tacoma Narrows Bridge in Washington, USA: (Left) opened on July 1, 1940 and (Right) collpased on Novermber 7, 1940 [23, 24]. (b) Hanshin Expressway bridge in Kobe, Japan: (Left) before and (Right) after collapse on January 18, 1995 [25, 26]. (c) I-35W Bridge in Minneapolis, USA: (Left) before and (Right) after collapse on August 1, 2007 [27]. (d) Morandi bridge in Genoa, Italy: (Left) before and (Right) after collapse on August 14th 2018 [28, 29]. (e) Nan Fang'ao bridge in Taiwan: (Left) before and (Right) after collapse on October 1, 2019 [30].*

## 4. Structural health monitoring (SHM)

Engineering assets such as high-rise buildings, long-span bridges, dams, oil platforms, hydraulic structures, wind turbines, transmission towers, ships, offshore structures, aircraft, and rail tracks may experience damage during

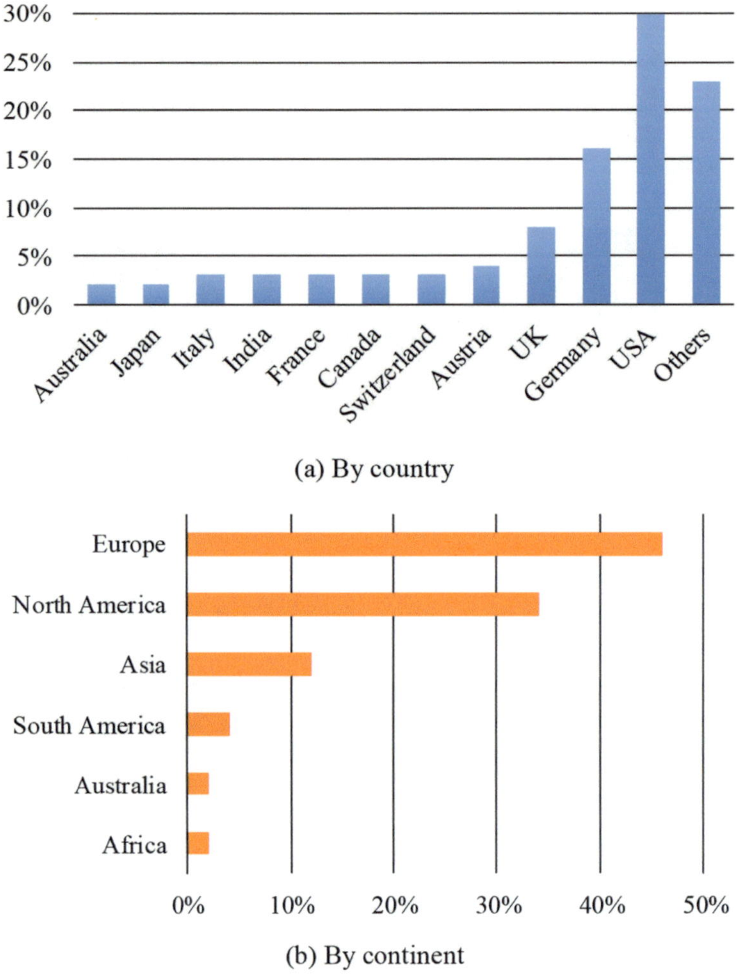


**Figure 3.**
*Geographical origin of recorded bridge failures. (a) By country. (b) By continent.*

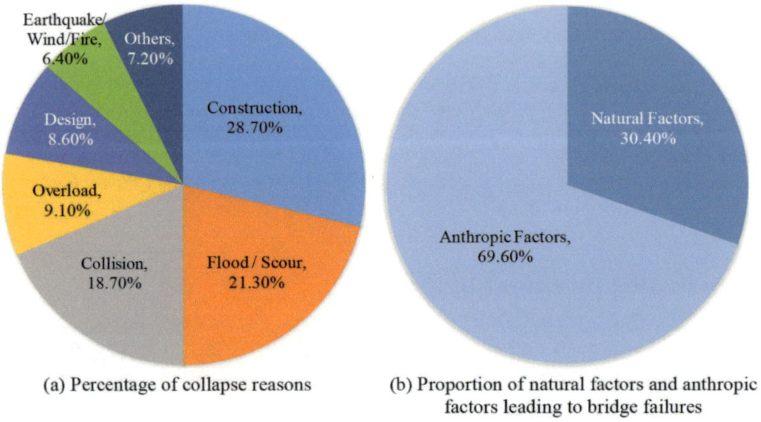


**Figure 4.**
*The distribution of bridge failure causes between 2009 and 2018. (a) Percentage of collapse reasons. (b) Proportion of natural factors and anthropic factors leading to bridge failures.*

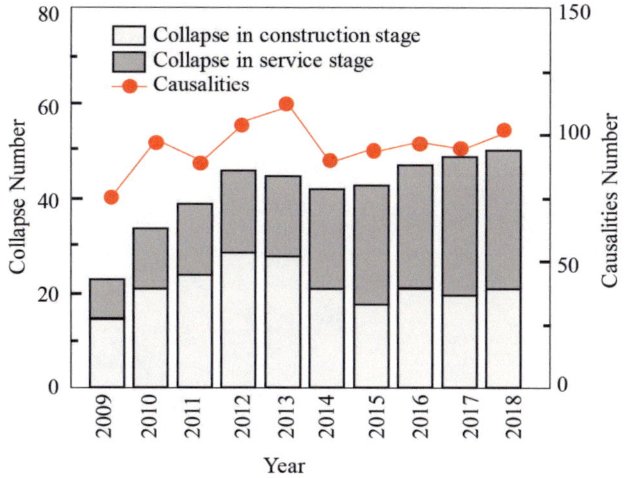


**Figure 5.**
*Number of bridge collapses and casualties between 2009 and 2018 [31].*

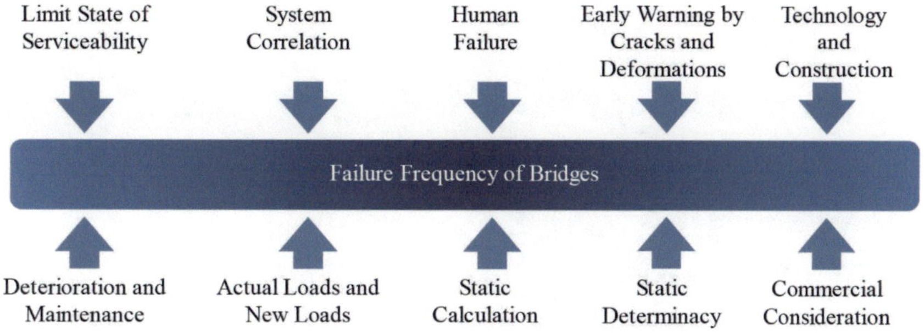


**Figure 6.**
*Potential factors influencing the observed failure frequency of bridges.*

construction or while in service induced by different reasons such as common weakening of material properties, fatigue, aging, delamination, wear, corrosion, creep, environmental effects (e.g., microstructural defects, cracking, thermal stress, residual stress, instability, and fastening or adhesive faults), overloading, changes in loading patterns or various unexpected causes such as wind excitations, earthquake, vehicle impact or blast loads during their service life, which can critically disturb their integrity, serviceability, and safety [32, 33]. Therefore, the unpredicted structural failure in such assets can produce catastrophic collapse, economic costs, human injuries, and death.

SHM is a transdisciplinary area of engineering applied to guarantee the operational safety and structural integrity of the materials, different components, or whole structure [34]. Engineers define health monitoring as the measurement of the operating and loading environment and the critical responses of a structure in order to track and evaluate the symptoms of operational anomalies and deterioration or damage that may impact service or safety reliability. The functionality of SHM systems is extremely similar to the human nervous system, as shown in **Figure 8**. The nervous system of humans consists of a complex collection of nerves and specialized cells and the main processing unit (brain). The nerve cells transmit signals between different

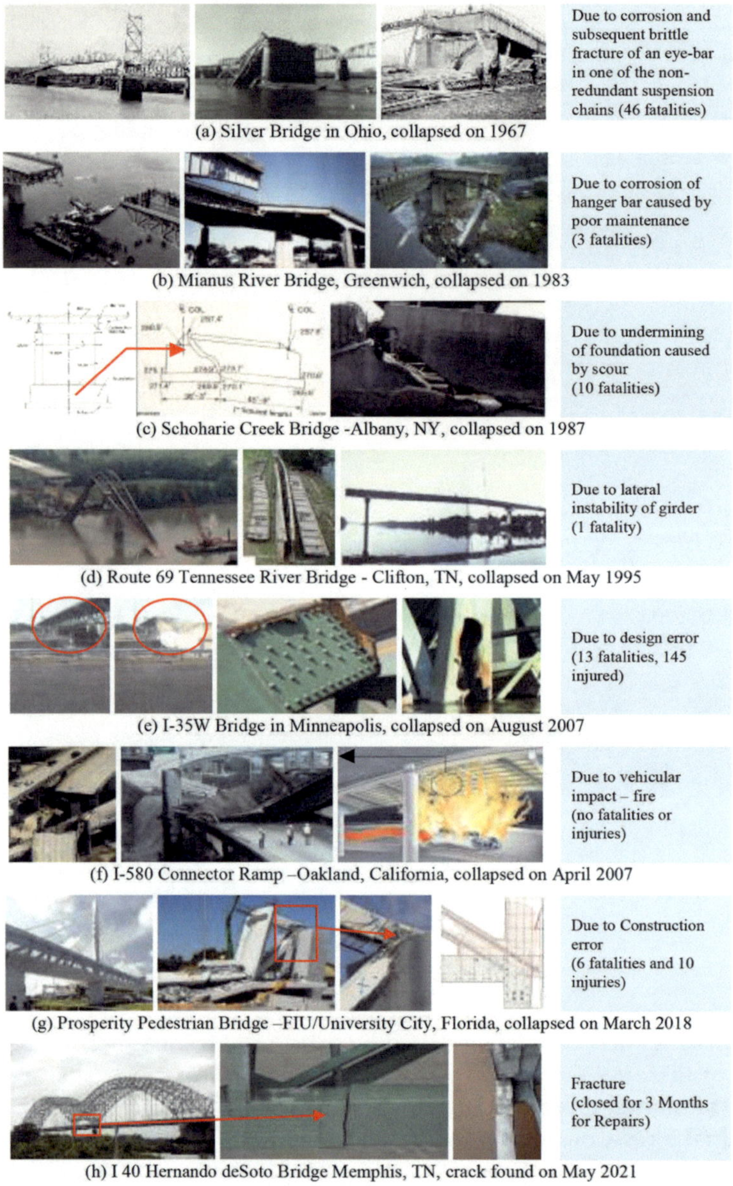


**Figure 7.**
*Causes of catastrophic bridge failures in the USA. (a) Silver Bridge in Ohio, collapsed on 1967. (b) Mianus River Bridge, Greenwich, collapsed on 1983. (c) Schoharie Creek Bridge—Albany, NY, collapsed on 1987. (d) Route 69 Tennessee River Bridge—Clifton, TN, collapsed on May 1995. (e) I-35W Bridge in Minneapolis, collapsed on August 2007. (f) I-580 Connector Ramp –Oakland, California, collapsed on April 2007. (g) Prosperity Pedestrian Bridge—FIU/University City, Florida, collapsed on March 2018. (h) I 40 Hernando deSoto Bridge Memphis, TN, crack found on May 2021.*

parts of the body and the brain. The brain is the main control unit for receiving and processing information as well as issuing instructions. In the same manner, SHM consists of a sensory network to gather information and a control unit for data processing and decision making [35]. Broadly speaking, the aims of conducting an SHM system

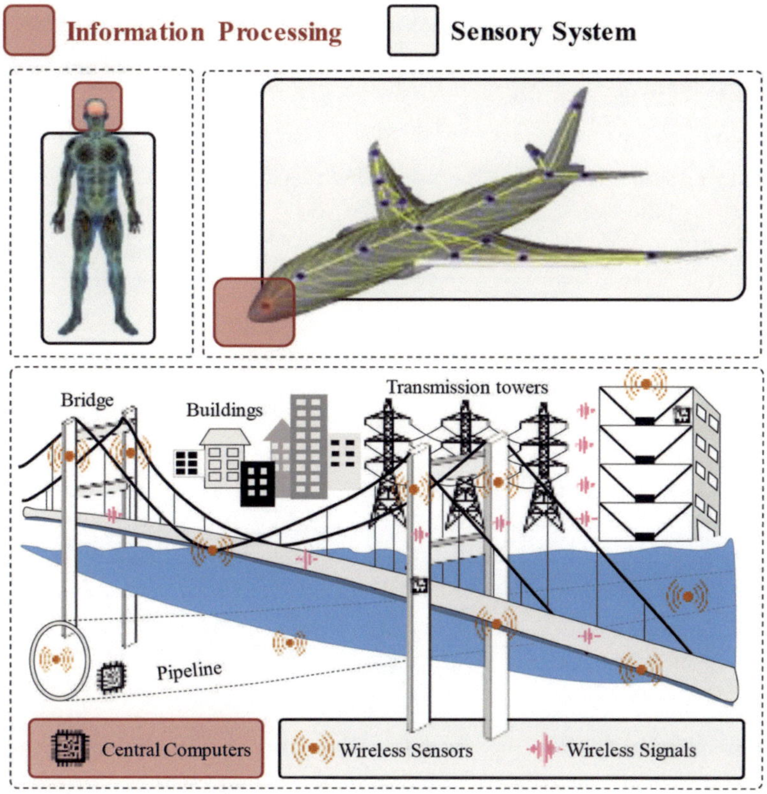


**Figure 8.**
*Similarity of SHM and human nervous system.*

are as follows: (1) To determine the current condition, (2) To predict future behavior, and (3) To early detect the structural damages.

The first bridge health monitoring was conducted in 1937 on the Golden Gate Bridge in San Francisco. Likewise, the direct application of data mining in structural damage identification was started in 2014 [36]. An overview of SHM and soft computing evolution over the years is shown in **Figure 9**. Eventually, the monitoring process in SHM generates massive data. Hence, precious information has to be acquired from unprocessed datasets [37]. However, data analysis of the generated big data obtained from the sensor network is a challenging task [36, 38]. To overcome this, data mining can therefore be employed to develop damage detection schemes [39, 40]. In computer science, data mining has become a research hotspot in recent years [41]. Consequently, data mining is offering a bright future ahead for other inquiries in various areas, such as aerospace, civil, industrial, and mechanical engineering. It is because data mining has a key role in the extraction of valuable information from different databases [6, 42].

With demanding needs to generate a massive volume of datasets, there has been a revolution in measuring and monitoring systems in the 1990s such as the development of sophisticated signal technologies, wireless networks, optical sensors, and global positioning systems. In this direction, the growth in the number of sensors installed on several important bridges worldwide during the past 20 years is shown in **Figure 10** [43].

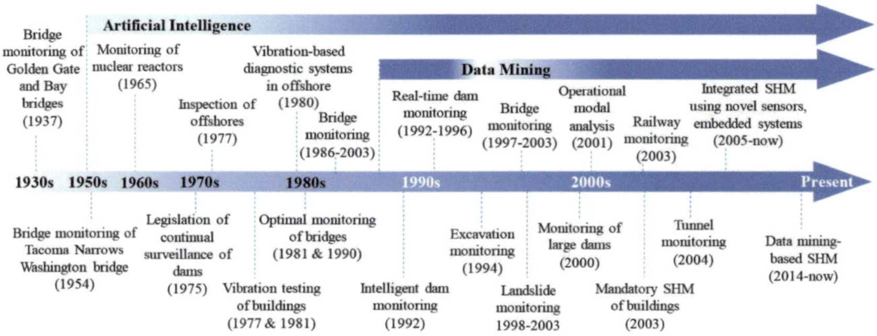


**Figure 9.**
*An overview of SHM and soft computing evolution over the years.*

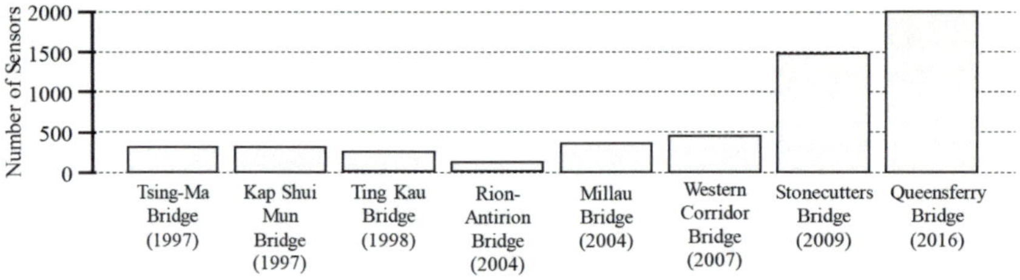


**Figure 10.**
*An example of the evolution of the number of sensors for bridge monitoring during the past 20 years, adopted from [43].*

## 5. Soft computing

Soft computing includes a series of strategies, that aim to exploit tolerance for imprecision, uncertainty, and partial truth to establish robustness and flexibility along with low solution cost. Major soft computing techniques and topics are summarized in **Figure 11**. In bridge monitoring, different soft computing techniques have been utilized for damage detection and system identification. The followings present the most commonly used soft computing applications for bridge monitoring.

Applications of data mining in SHM have recently been reported [44, 45], though due to the novelty of data mining, the application of data mining in SHM is still controversial and is not as much as expected. Therefore, it seems necessary studies are required to advance the data mining application in SHM. One of the most widespread systematic data mining tools is Cross Industry Standard Process for Data Mining (CRISP-DM), which was introduced by a consortium of several companies such as National Cash Register (NCR) System Engineering Copenhagen from the USA and Denmark, Integral Solutions Ltd. (ISL)/SPSS from the USA, Daimler Chrysler AG from Germany and an insurance corporation in the Netherlands, called OHRA [46, 47]. A generalized form of CRISP-DM based on the SHM system has been developed by the authors of [48]. The proposed data mining-based damage identification approach consisted of six new defined stages: target identification, data exploration, database construction, pattern identification, pattern evaluation, and knowledge extraction. At the first stage, specimen description and experimental setup have been presented using a slab-on-girder bridge structure (see **Figure 12**). In the second stage, vibration data were collected from experimental modal analysis of healthy (baseline) and damaged

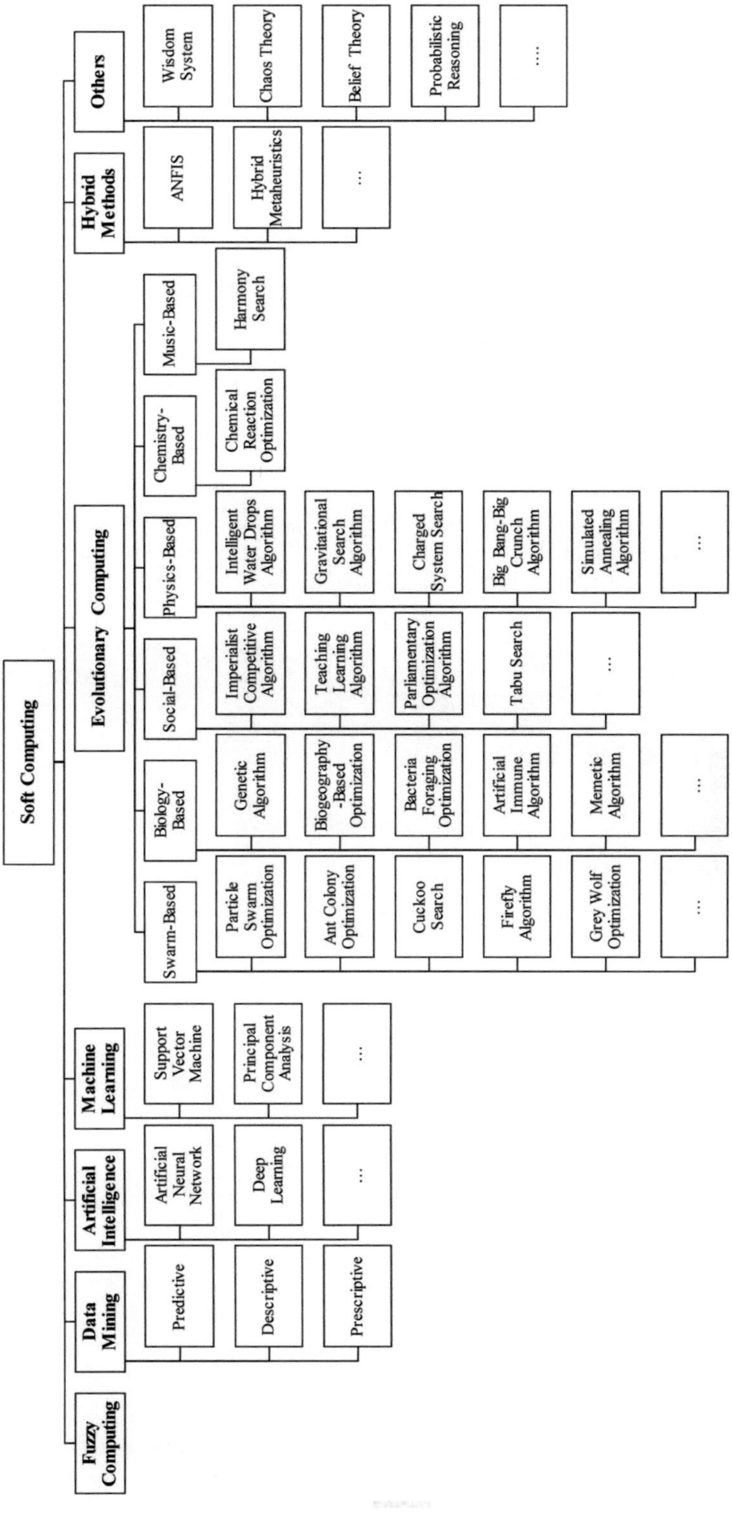


**Figure 11.**
*Soft computing techniques.*

structures. Analysis of collected data was done in the third stage to generate datasets using the first four flexural modes and all corresponding mode shape values of double-point damage cases as inputs for next stage, which was pattern identification. In this stage, Artificial Neural Network (ANN)-based Imperial Competitive Algorithm (ICA) was employed to train datasets and build a model for damage identification of the structure. Then, in the fifth stage, model performance was assessed using evaluation methods and ANN approach. Finally, the last stage extracted valuable knowledge and damage identification.

A multi-layer perceptron ANN-based damage detection of truss bridge joints was proposed by Mehrjoo et al. [49]. The proposed network was conducted using a single hidden layer and 729 training patterns obtained from the first five modes of the structure. It should be noted that the stiffness reduction of members was considered as damage in this study. **Figure 13** illustrates the fatigue damage in truss bridge joints. Around 40% of these types of structures usually experience fatigue damage in their joints during their service life. Louisville bridge was used as a test specimen for this research. The standard back-propagation (BP) algorithm with 75,000 epochs and the root mean square (RMS) was employed to train the networks and evaluation of patterns, respectively. The average error of the predicted damage percentage value for all joints was 1.28%. The authors summarized that their presented methodology was effective in system identification of truss bridges.

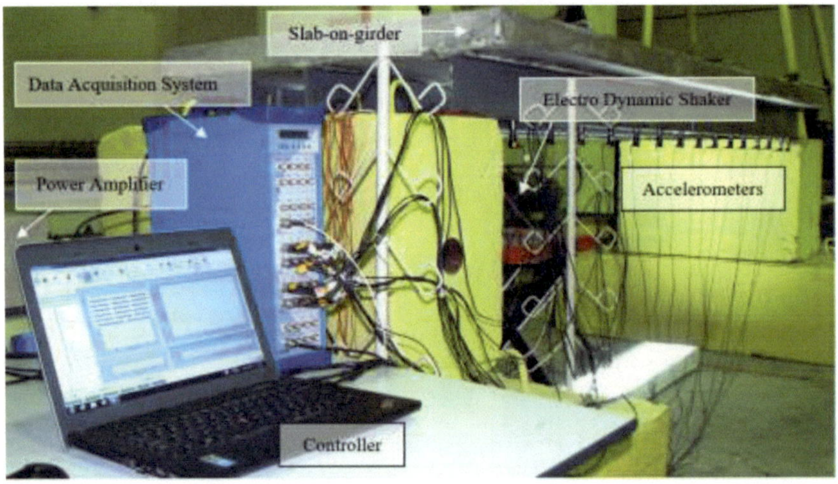


**Figure 12.**
*Experimental test of the lab-scale bridge [37].*

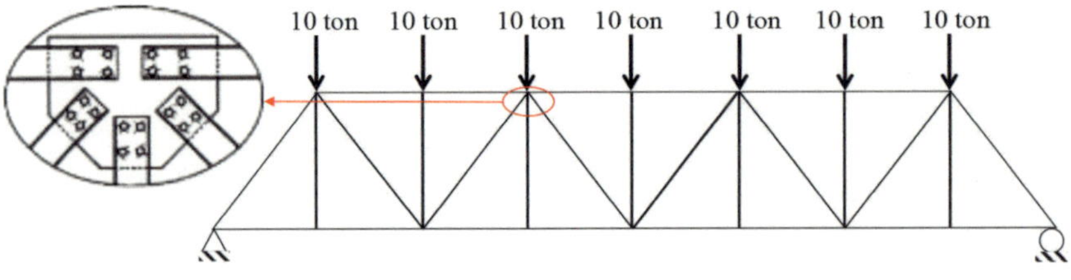


**Figure 13.**
*Fatigue damage in truss bridge joints in Louisville bridge truss [49].*

A bridge monitoring scheme using operational modal analysis was developed by [50] through a hybrid Fuzzy Krill Herd approach. The proposed fuzzy logic-based SHM diagram is displayed in **Figure 14a**. Two types of bridges, i.e., Banafjäl bridge in Sweden and the Tirehrood bridge in Iran, were considered as test specimens for this paper. The damage scenarios were presented by the output of the fuzzy logic-based SHM approach, as shown in **Figure 14b**. The outcomes revealed the proficiency of the proposed approach in achieving precise knowledge in the existence of noisy data.

Padil et al. [51] proposed a PCA-based non-probabilistic technique. The proposed method was verified using a big Frequency Response Function (FRF) matrix comprising 1200 FRFs with 512 frequency points obtained from intact and damaged simply-supported steel truss bridge model (see **Figure 15a**). To aid the aim, a number of damage scenarios were considered by cutting the structural members, i.e., M1 and

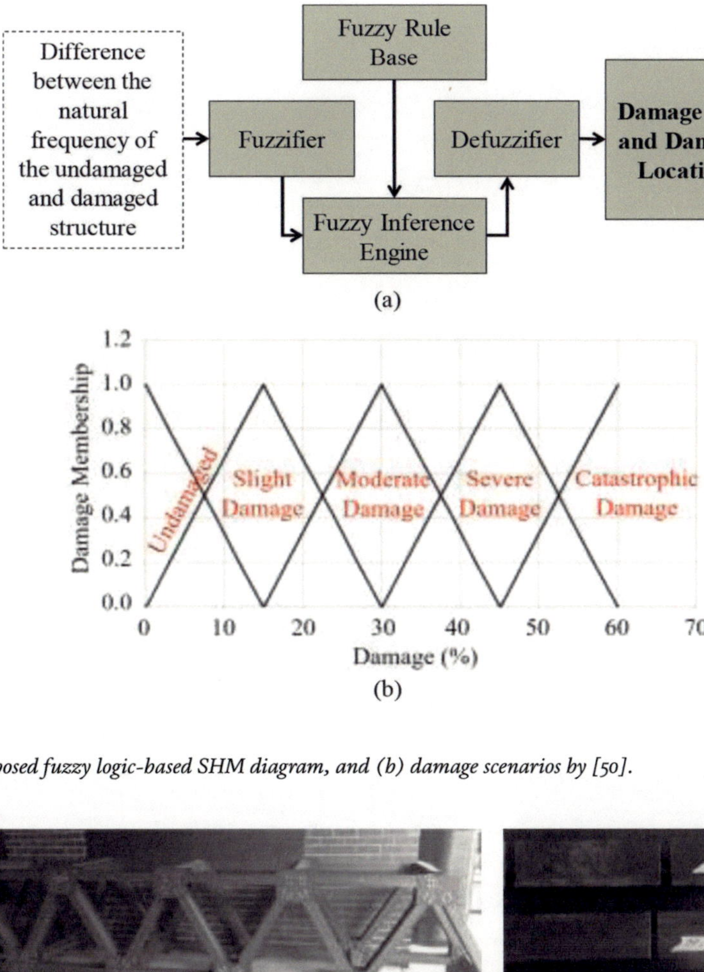


**Figure 14.**
*(a) The proposed fuzzy logic-based SHM diagram, and (b) damage scenarios by [50].*

**Figure 15.**
*(a) Experimental model of a steel truss bridge and (b) imposed damages [51].*

M2 in the main girder bar, and W1 in the web bar, as shown in **Figure 15b**. In this study, it was shown that the results of hybrid PCA were better than traditional PCA.

The application of deep learning in SHM was implemented in [52] by different PCA-based methods, i.e., the deep principal component analysis (DPCA), nonlinear principal component analysis (NLPCA), and kernel principal component analysis (KPCA). The damage-sensitive features of Z24 and Tamar Bridges were used to evaluate the applicability of the aforementioned algorithms (see **Figure 16**). The DPCA showed the best performance.

**Figure 16.**
*(a) Z24 bridge, and (d) Tamar suspension bridge [52].*

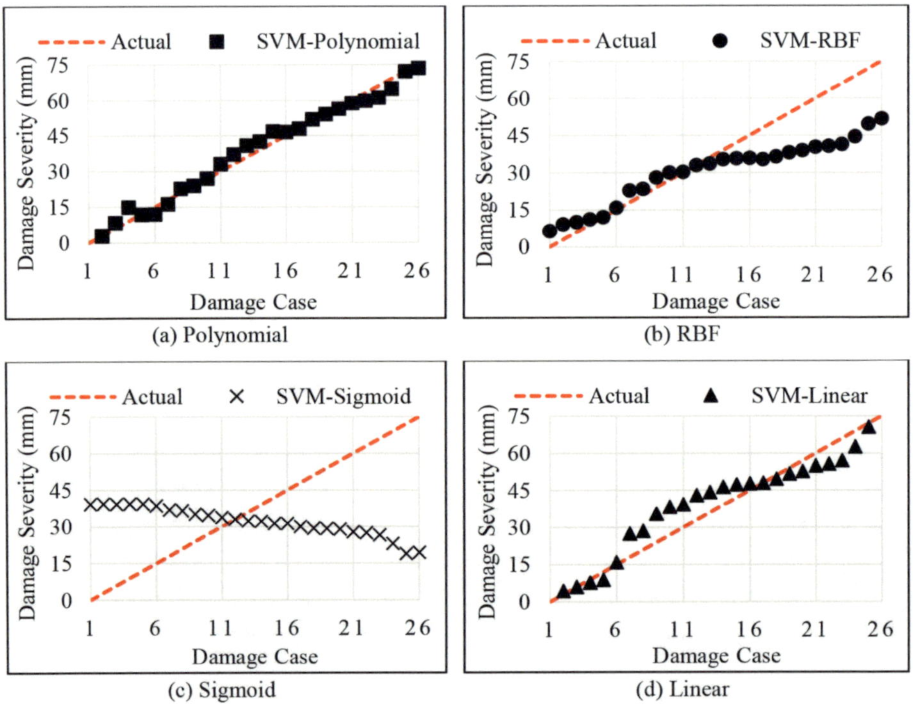


**Figure 17.**
*SVM results for composite bridge structure using different kernel functions [36].*

In [36], an SVM model was carried out using four different kernel functions including the Gaussian radial basis function (RBF), Polynomial, Sigmoid, and Linear kernel functions. Experimental modal analysis of a bridge structure was performed to generate the modal parameters as the input database for the model creation. A number of damage cases were conducted to predict the damage severity. As shown in **Figure 17**, amongst all patterns, SVM-Polynomial achieved the most accurate predicted outputs. To offer an explanation, kernel functions were used in order to bring the data from a lower dimension to a higher dimension. To this end, SVM classifier divided the data with a new plane, i.e., hyperplane. Therefore, despite the better learning power in RBF kernel amongst others, this local function could not provide a satisfying dissemination efficiency. Instead, the polynomial kernel, which is a global function, performed a superior data dissemination strategy. Nonetheless, the learning process of polynomial function experienced a lower level of learning capacity.

A finite element model reduction methodology using a Bayesian inference was developed in [53] for structural bolted-connection damage detection. A novel likelihood-free Bayesian inference method for structural parameter identification has also been proposed in [54]. To aid the aim, the numerical simulations of a three-dimensional bridge structure were carried out to validate the performance and accuracy of the proposed approach (see **Figure 18**). It was concluded that the reported Bayesian model updating technique was capable of predicting the posterior probabilities of unknown structural parameters.

**Figure 18.**
*Finite element model of a bridge structure [54].*

**Figure 19.**
*Nam O railway bridge [55].*

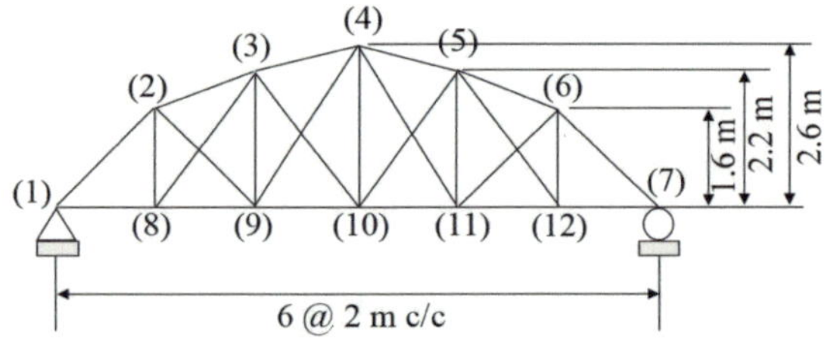


E=200 Gpa ρ=7800 kg/m$^3$

| Member | 1-6 | 7-12 | 13-17 | 18-25 |
|---|---|---|---|---|
| Area (mm$^2$) | 1800 | 1500 | 1000 | 1200 |

**Figure 20.**
*25-member plane truss [58].*

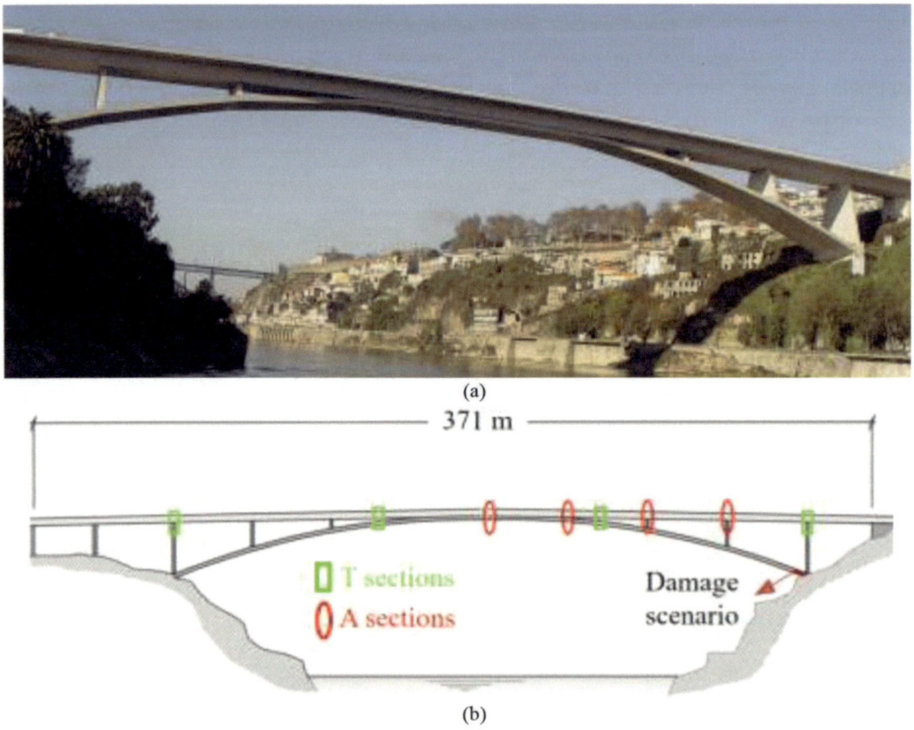


**Figure 21.**
*SHM of the infante D. Henrique bridge (a) bridge image, and (b) sensors locations (a and T represent acceleration and temperature, respectively) [60].*

The Nam O Railway Bridge is a large-scale steel truss bridge located on the unique main rail track from the north to the south of Vietnam (see **Figure 19**). In [55], the experimental measurements obtained from this bridge were carried out under ambient vibrations using piezoelectric sensors, and a finite element model was also created in MATLAB to represent the physical behavior of the structure. By model updating, the discrepancies between the experimental and the numerical results were minimized. For the success of the model updating, the efficiency of the optimization

| Application | Algorithm | References |
|---|---|---|
| Faulty data detection | Convolutional Neural Network (CNN) | [61] |
| Shear loading detection | Memory-Augmented CNN | [62] |
| Concrete crack detection | Faster region-based CNN | [63] |
| Data anomaly detection | One-dimensional CNN | [64] |
| Rust grade recognition | Ensemble CNN with voting strategy | [65] |
| Data anomaly detection | Deep Neural Network (DNN) | [66] |
| Automated crack evaluation | Deep learning-based segmentation | [67] |
| Feature selection | ANN | [68] |
| Fatigue damage detection | ANN | [69] |
| Probabilistic Damage detection | ANN with Bayesian | [70] |
| Pattern recognition | ANN | [71] |
| On-line early-damage detection | ANN | [72] |
| Damage detection | Extended Kalman filter-based ANN | [73] |
| Damage quantification | ANN | [74] |
| Temperature effect removal | Auto-associative ANN | [75] |
| Damage detection | Auto-associative ANN | [76] |
| Damage classification | ANN | [77] |
| Continuous online damage detection | PCA | [78] |
| Bridge health monitoring | Robust PCA | [79] |
| Damage identification | Double-window PCA (DWPCA) | [80] |
| Data-driven damage detection | Fixed moving PCA | [81] |
| Nonparametric damage detection | PCA | [82] |
| Missing data estimation in SHM | Probabilistic PCA | [83] |

**Table 1.**
*Summary of bridge monitoring using advanced computational techniques.*

algorithm was essential. Particle swarm optimization (PSO) algorithm and genetic algorithm (GA) were employed to update the unknown model parameters. The authors claimed that not only the PSO result showed better accuracy but also reduced the computational cost compared to GA. This study focused on the stiffness conditions of typical joints of truss structures. According to their results, the assumption of semi-rigid joints (using rotational springs) could most accurately represent the dynamic characteristics of the truss bridge.

A damage assessment based on ACO was proposed in [56] from changes in natural frequencies. A plane truss structure was considered in this study to validate the efficiency and robustness of the presented methodology. The authors reported that their method was capable of correct damage assessment even with a noisy dataset. An inverse vibration-based approach using ACO and natural frequencies changes has also been carried out in [57]. The authors of this research made a comparison between ACO and PSO. Furthermore, the performance of a continuous ACO and PSO was also evaluated in [58] for damage detection of plane and space truss structures based on frequency and mode shapes-based objective function. The details of the truss structure are displayed in **Figure 20**.

A regression-based damage detection approach was developed by [59] using the natural frequencies of the Z24 Bridge. In this research, the traditional regression, as well as the developed regression, was applied to the illustrative structures for identifying the existence of damage. It was established that both methods could detect the presence of damage. However, better outcomes were acquired by the developed regression-based damage detection approach. Another regression model has been developed by [60] as an up-to-date damage detection scheme in a field-monitored bridge. To aid the aim, the Infante D. Henrique Bridge in Portugal has been continuously monitored since 2007 using two synchronized Global Positioning System (GPS)-based data analyzers, temperature and vibration accelerometers (see **Figure 21**). It is worth noting that the recorded data were less contaminated by noise through GPS connection. The measured natural frequencies change corresponding to the damage scenarios by reduction in the vertical bending inertia of the arch were employed in this work to obtain damage-sensitive features.

**Table 1** summarizes the latest applications of artificial intelligence methods in bridge health monitoring.

## 6. Conclusion

A bridge is one of the most symbolic, important, as well as expressive infrastructures worldwide for social and economic activities of mankind where it serves as the crucial link in the transport network. Therefore, condition assessment and damage detection of this asset is frequently required to guarantee the safe functioning of the infrastructure. To do so, SHM systems have been applied to make satisfactory decisions on structural maintenance, repair, and rehabilitation. However, conventional SHM cannot be used for structural continuous monitoring, real-time and online assessment to solve real-world problems. Therefore, integration of SHM with soft computing techniques has been successfully applied for optimized monitoring of bridges in recent years. This is due to the fact that soft computing is an umbrella of computational techniques that tolerates uncertainty imprecision, partial truth, and ambiguity. Hence, this chapter introduced the optimized SHM-based soft computing techniques of bridge structures through artificial intelligence and machine learning algorithms in order to illustrate the performance of advanced bridge monitoring approaches, which were required to maintain the health condition of infrastructures and for smooth functioning of cities.

## Acknowledgements

The authors wish to acknowledge the University of Malaya and K.N.TOOSI University of Technology for providing the resources and supporting this research. The authors would also like to express their sincere thanks to the Structural Health Monitoring Research Group (StrucHMRSGroup), which was led by Professor Emeritus Hashim Abdul Razak. (Program Number: IIRG007A-2019).

## Author details

Meisam Gordan[1,2]*, Saeed-Reza Sabbagh-Yazdi[2], Khaled Ghaedi[1], David P. Thambiratnam[3] and Zubaidah Ismail[1]

1 Department of Civil Engineering, University of Malaya, Kuala Lumpur, Malaysia

2 Department of Civil Engineering, K.N. TOOSI University of Technology, Tehran, Iran

3 School of Civil Engineering and Built Environment, Queensland University of Technology (QUT), Brisbane, QLD, Australia

*Address all correspondence to: meisam.gordan@gmail.com

## References

[1] Wang T, Bhuiyan MZA, Wang G, Rahman MA, Wu J, Cao J. Big data reduction for a Smart City's critical infrastructural health monitoring. IEEE Communications Magazine. 2018;**56**:128-133. DOI: 10.1109/MCOM.2018.1700303

[2] Khan AA, Zafar S, Khan NS, Mehmood Z. History, current status and challenges to structural health monitoring system aviation field. JavaServer Pages Technology. 2014;**4**:67-74

[3] Di Sante R. Fibre optic sensors for structural health monitoring of aircraft composite structures: Recent advances and applications. Sensors (Switzerland). 2015;**15**:18666-18713. DOI: 10.3390/s150818666

[4] Reagan D, Sabato A, Niezrecki C. Feasibility of using digital image correlation for unmanned aerial vehicle structural health monitoring of bridges. Structural Control and Health Monitoring. 2018;**17**:1056-1072. DOI: 10.1177/1475921717735326

[5] Bessa MA. Data-Driven Multi-Scale Analyses of Materials and Structures, PhD Dissertation. Illinos: Northwestern University; 2016

[6] Gordan M, Ismail ZB, Razak HA, Ghaedi K. Optimization-based evolutionary data mining techniques for structural health monitoring. Journal of Civil Engineering and Construction. 2020;**9**:14-23

[7] Heggade VN. The conceptual Design of Bridges: Form finding and aesthetics. Structural Engineering International. 2021;**31**:622-637. DOI: 10.1080/10168664.2021.1899780

[8] Carfrae T, Bardsley H, Lenczner A. Eminent structural engineer: Peter rice. Structural Engineering International. 2018;**28**:556-559. DOI: 10.1080/10168664.2018.1498302

[9] Hansford M. Winning the emotional argument. New Civil Engineer. 2017;**25**:e859 Available from: https://www.newcivilengineer.com/archive/comment-winning-the-emotional-argument-06-02-2017/

[10] Knight M. BIM and the art of motorcycle maintenance. Structural Engineering International. 2018;**28**:457-461. DOI: 10.1080/10168664.2018.1468230

[11] Beade-Pereda H. The responsibility of the bridge designer. Structural Engineering International. 2017;**27**:337. DOI: 10.1080/10168664.2017.11985640

[12] Terms F. IABSE news. Structural Engineering International. 2017;**27**:140-151. DOI: 10.1080/10168664.2017.11985618

[13] Youssef M. Problems of neglected places under bridges: A case study of Yerevan bridge, Beirut, Lebanon. WIT Transactions on Ecology and the Environment. 2017;**226**:739-750. DOI: 10.2495/SDP170641

[14] Arslan A. Bridges as city landmarks: A critical review on iconic structures. Journal of Design Studio. 2020;**2**:85-99. DOI: 10.46474/jds.798072

[15] Sani RM. A conceptual understanding for teaching the history of Islamic architecture: An Iranian (Persian) Perspective. International Journal of Architectural Research (IJAR). 2009;**3**:233-244. DOI: 10.26687/archnet-ijar.v3i1.266

[16] Knapp RG, Miller TE, Liu J. China's corridor bridges: heritage buildings over

water. Built Heritage. 2020;**4**:e456. DOI: 10.1186/s43238-020-00010-w

[17] Gauvreau P. Educating engineers to create good looking bridges. Structural Engineering International. 2016;**26**:198-206. DOI: 10.2749/101686616X14555428759406

[18] Jákli A. 14th International Conference on Ferroelectric Liquid Crystals, Magdeburg, Germany. Magdeburg, Germany: IEEE; 2014. DOI: 10.1080/1358314X.2014.898871

[19] Arjmand M, Bratek W. Operational modal analysis to identify modal parameters in reciprocating compressors. In: GMRC Gas Machinery Conference. San Antonio, TX: IEEE; 2019

[20] Mulqueen PC. Creating the Te Rewa Rewa bridge, New Zealand. Structural Engineering International: Journal of the International Association for Bridge and Structural Engineering. 2011;**21**:486-491. DOI: 10.2749/101686611X13131377726126

[21] Siwowski T, Wysocki A. The circular footbridge of Rzeszów, Poland. Structural Engineering International: Journal of the International Association for Bridge and Structural Engineering. 2015;**25**:460-466. DOI: 10.2749/101686615X14355644771018

[22] Sander C. Coming to light: Principles for successful lighting design of footbridges. Structural Engineering International. 2018;**28**:448-456. DOI: 10.1080/10168664.2018.1488555

[23] Washington State Department of Transportation (WSDOT), Tacoma Narrows Bridge history. https://www.wsdot.wa.gov/TNBhistory/

[24] Plaut RH. Snap loads and torsional oscillations of the original Tacoma narrows bridge. Journal of Sound and Vibration. 2008;**309**:613-636. DOI: 10.1016/j.jsv.2007.07.057

[25] Venton D. Tectonic tremors could offer insights into the big shakers. Proceedings of the National Academy of Sciences of the United States of America. 2016;**113**:7930-7931. DOI: 10.1073/pnas.1610000113

[26] Hayashi H, Marui T, Taniguchi N, Kayano S. Restoration of Hanshin expressway after Kobe/Awaji earthquake - challenge of 623 days before opening. Cement and Concrete Composites. 2000;**22**:29-38. DOI: 10.1016/S0958-9465(99)00046-3

[27] Swartz RA, Zimmerman A, Lynch JP. Structural health monitoring system with the latest information technologies. In: Proceedings of 5th Infrastructure & Environmental Management Symposium. Yamaguchi, Japan: IEEE; 2007

[28] Russell C. Before and after photos of Genoa bridge collapse, Stuff Limited. 2018. Available from: https://www.stuff.co.nz/world/ europe/106302828/before-and-after-photos-of-genoa-bridge-collapse

[29] Milillo P, Giardina G, Perissin D, Milillo G, Coletta A, Terranova C. Pre-collapse space geodetic observations of critical infrastructure: The Morandi Bridge, Genoa, Italy. Remote Sensor. 2019;**11**:e457. DOI: 10.3390/rs11121403

[30] Horgan R. Fatal Taiwan bridge collapse is latest example of maintenance failings, New Civil Engineer. 2019. Available from: https://www.newcivilengineer.com/ latest/fatal-taiwan-bridge-collapse-

[31] Tan JS, Elbaz K, Wang ZF, Shen JS, Chen J. Lessons learnt from bridge collapse: A view of sustainable management. Sustain. 2020;**12**:1-16. DOI: 10.3390/su12031205

[32] Mccrory JP, Al-jumaili SK, Crivelli D, Pearson MR, Eaton MJ, Featherston CA, et al. Damage classification in carbon fibre composites using acoustic emission: A comparison of three techniques. Composites: Part B. 2015;**68**:424-430. DOI: 10.1016/j.compositesb.2014.08.046

[33] Seyedpoor SM, Shahbandeh S, Yazdanpanah O. An efficient method for structural damage detection using a differential evolution algorithm-based optimisation approach. Civil Engineering and Environmental Systems. 2015;**32**:230-250. DOI: 10.1080/10286608.2015.1046051

[34] Yuan F-G, Zargar SA, Chen Q, Wang S. Machine learning for structural health monitoring: challenges and opportunities. Sensors and Smart Structures Technologies for Civil, Mechanical, and Aerospace Systems. 2020;**113**:1137903. DOI: 10.1117/12.2561610

[35] Abdo MA-B. Structural Health Monitoring History, Applications and Future, A review book. Paris, France: Open Science; 2015

[36] Gordan M. Data mining for structural damage identification using hybrid artificial neural network based algorithm for beam and slab girder [PhD thesis]. Kuala Lumpur, Malaysia: University of Malaya; 2020

[37] Gordan M, Ismail Z, Abdul Razak H, Ghaedi K, Ibrahim Z, Tan ZX, et al. Data mining-based damage identification of a slab-on-girder bridge using inverse analysis. Measurement. 2020;**151**:107175. DOI: 10.1016/j.measurement.2019.107175

[38] Talebkhah M, Sali A, Marjani M, Gordan M, Hashim SJ, Rokhani FZ. Edge computing: architecture, applications and future perspectives. In: IICAIET2020 (IEEE International Conference on Artificial Intelligence in Engineering and Technology). Sabah, Malaysia: IEEE; 2020. DOI: 10.1109/IICAIET49801.2020.9257824

[39] Gordan M, Razak HA, Ismail Z, Ghaedi K. Data mining based damage identification using imperialist competitive algorithm and artificial neural network. Latin American Journal of Solids and Structures. 2018;**15**:1-14. DOI: 10.1590/1679-78254546

[40] Gordan M, Ismail Z, Razak HA, Ibrahim Z. Vibration-based structural damage identification using data mining. In: 24th International Congress on Sound and Vibration. London: ICSV; 2017

[41] Lai Y, Magd Z, Wahab A, Maia NMM, Liu L. Data Mining in Structural Dynamic Analysis. Berlin, Germany: Springer Nature; 2019. DOI: 10.1007/978-981-15-0501-0

[42] Gordan M, Ismail Z, Ibrahim Z, Hashim H. Data mining technology for structural control systems: concept, development, and comparison. In: Recent Trends Artificial Neural Networks. London: IntechOpen Limited; 2019. DOI: 10.5772/intechopen.88651

[43] Cremona C, Santos J. Structural health monitoring as a big-data problem structural health monitoring as a big-data problem. Structural Engineering International. 2018;**28**:243-254. DOI: 10.1080/00000000.2012.000000

[44] Gordan M, Razak HA, Ismail Z, Ghaedi K. Recent developments in damage identification of structures using data mining. Latin American Journal of Solids and Structures. 2017;**14**:2373-2401

[45] Li X, Yu W, Villegas S. Structural health monitoring of building structures

with online DataMiningMethods. IEEE Systems Journal. 2016;**10**:1291-1300

[46] Chapman P, Clinton J, Kerber R, Khabaza T, Reinartz T, Shearer C, et al. CRISP-DM 1.0 Step-by-step data mininng guide. SPSS Inc. 2000;**9**:13

[47] Fernandez IB, Zanakis SH, Walczak S. Knowledge discovery techniques for predicting country investment risk. Computers and Industrial Engineering. 2002;**43**:787-800

[48] Gordan M, Razak HA, Ismail Z, Ghaedi K, Tan ZX, Ghayeb HH. A hybrid ANN-based imperial competitive algorithm methodology for structural damage identification of slab-on-girder bridge using data mining. Applied Soft Computing Journal. 2020;**88**:106013. DOI: 10.1016/j.asoc.2019.106013

[49] Mehrjoo M, Khaji N, Moharrami H, Bahreininejad A. Damage detection of truss bridge joints using artificial neural networks. Expert Systems with Applications. 2008;**35**:1122-1131. DOI: 10.1016/j.eswa.2007.08.008

[50] Jahan S, Mojtahedi A, Mohammadyzadeh S, Hokmabady H. A fuzzy Krill Herd approach for structural health monitoring of bridges using operational modal analysis, Iran. Iranian Journal of Science and Technology, Transactions of Civil Engineering. 2020;**45**(2):1139-1157. DOI: 10.1007/s40996-020-00475-w

[51] Padil KH, Bakhary N, Abdulkareem M, Li J, Hao H. Non-probabilistic method to consider uncertainties in frequency response function for vibration-based damage detection using artificial neural network. Journal of Sound and Vibration. 2020;**467**:115069. DOI: 10.1016/j.jsv.2019.115069

[52] Silva M, Santos A, Santos R, Figueiredo E, Sales C, Costa JCWA. Deep principal component analysis: An enhanced approach for structural damage identification. Structural Control and Health Monitoring. 2019;**18**:1444-1463. DOI: 10.1177/1475921718799070

[53] Yin T, Jiang Q, Yuen K. Vibration-based damage detection for structural connections using incomplete modal data by Bayesian approach and model reduction technique. Engineering Structures. 2017;**132**:260-277

[54] Ni P, Han Q, Du X, Cheng X. Bayesian model updating of civil structures with likelihood-free inference approach and response reconstruction technique. Mechanical Systems and Signal Processing. 2022;**164**:108204. DOI: 10.1016/j.ymssp.2021.108204

[55] Tran-Ngoc H, Khatir S, De Roeck G, Bui-Tien T, Nguyen-Ngoc L, Abdel Wahab M. Model updating for Nam O bridge using particle swarm optimization algorithm and genetic algorithm. Sensors (Switzerland). 2018;**18**:1431. DOI: 10.3390/s18124131

[56] Majumdar A, Kumar D, Maity D. Damage assessment of truss structures from changes in natural frequencies using ant colony optimization. Applied Mathematics and Computation. 2012;**218**:9759-9772

[57] Majumdar A, Nanda B. A comparative study on inverse vibration based damage assessment techniques in beam structure using ant Colony optimization and particle swarm optimization. Advanced Science, Engineering and Medicine. 2020;**12**:918-923

[58] Barman SK, Maiti DK, Maity D. Damage detection of truss employing

swarm-based optimization techniques: A comparison. In: Advanced Engineering Optimization Through Intelligent Techniques. Singapore: Springer; 2020. pp. 21-37. DOI: 10.1007/978-981-13-8196-6

[59] Wah WSL, Chen YT, Owen JS. A regression-based damage detection method for structures subjected to changing environmental and operational conditions. Engineering Structures. 2021;**228**:111462. DOI: 10.1016/j.engstruct.2020.111462

[60] Comanducci G, Magalhães F, Ubertini F, Cunha Á. On vibration-based damage detection by multivariate statistical techniques: Application to a long-span arch bridge. Structural Control and Health Monitoring. 2016;**15**:505-524

[61] Jian X, Zhong H, Xia Y, Sun L. Faulty data detection and classification for bridge structural health monitoring via statistical and deep-learning approach. Structural Control and Health Monitoring. 2021;**28**:e2824. DOI: 10.1002/stc.2824

[62] Wang F, Song G, Mo YL. Shear loading detection of through bolts in bridge structures using a percussion-based one-dimensional memory-augmented convolutional neural network. Computer-Aided Civil and Infrastructure Engineering. 2021;**36**:289-301. DOI: 10.1111/mice.12602

[63] Deng J, Lu Y, Lee VCS. Concrete crack detection with handwriting script interferences using faster region-based convolutional neural network. Computer-Aided Civil and Infrastructure Engineering. 2020;**35**:373-388. DOI: 10.1111/mice.12497

[64] Ni FT, Zhang J, Noori MN. Deep learning for data anomaly detection and data compression of a long-span suspension bridge. Computer-Aided Civil and Infrastructure Engineering. 2020;**35**:685-700. DOI: 10.1111/mice.12528

[65] Xu J, Gui C, Han Q. Recognition of rust grade and rust ratio of steel structures based on ensembled convolutional neural network. Computer-Aided Civil and Infrastructure Engineering. 2020;**35**:1160-1174. DOI: 10.1111/mice.12563

[66] Bao Y, Tang Z, Li H, Zhang Y. Computer vision and deep learning–based data anomaly detection method for structural health monitoring. Structural Control and Health Monitoring. 2019;**18**:401-421. DOI: 10.1177/1475921718757405

[67] Jang K, An YK, Kim B, Cho S. Automated crack evaluation of a high-rise bridge pier using a ring-type climbing robot. Computer-Aided Civil and Infrastructure Engineering. 2021;**36**:14-29. DOI: 10.1111/mice.12550

[68] Okazaki Y, Okazaki S, Asamoto S, Chun PJ. Applicability of machine learning to a crack model in concrete bridges. Computer-Aided Civil and Infrastructure Engineering. 2020;**35**:775-792. DOI: 10.1111/mice.12532

[69] Rageh A, Eftekhar Azam S, Linzell DG. Steel railway bridge fatigue damage detection using numerical models and machine learning: Mitigating influence of modeling uncertainty. International Journal of Fatigue. 2020;**134**:105458. DOI: 10.1016/j.ijfatigue.2019.105458

[70] Yin T, Zhu HP. Probabilistic damage detection of a steel truss bridge model by optimally designed bayesian neural network. Sensors (Switzerland). 2018;**18**:3371. DOI: 10.3390/s18103371

[71] Santos J, Crémona C, Calado L. Real-time damage detection based on

pattern recognition. Structural Concrete. 2016;**17**:338-354

[72] Santos JP, Cremona C, Calado L, Silveira P, Orcesi AD. On-line unsupervised detection of early damage. Structural Control and Health Monitoring. 2016;**23**:1047-1069

[73] Jin C, Jang S, Sun X, Li J, Christenson R. Damage detection of a highway bridge under severe temperature changes using extended Kalman filter trained neural network. Journal of Civil Structural Health Monitoring. 2016;**6**:545-560. DOI: 10.1007/s13349-016-0173-8

[74] Chun P, Yamashita H, Furukawa S. Bridge damage severity quantification UsingMultipoint acceleration measurement and artificial neural networks. Shock and Vibration. 2015;**2015**:789384

[75] Zhou HF, Ni YQ, Ko JM. Eliminating temperature effect in vibration-based structural damage detection. Journal of Engineering Mechanics. 2011;**137**:785-797

[76] Hsu T, Loh C. Damage detection accommodating nonlinear environmental effects by nonlinear principal component analysis. Structural Control and Health Monitoring. 2010;**17**:338-354

[77] Kabir S, Rivard P, Ballivy G. Neural-network-based damage classification of bridge infrastructure using texture analysis. Canadian Journal of Civil Engineering. 2008;**35**:258-267

[78] Meixedo A, Santos J, Ribeiro D, Calçada R, Todd MD. Online unsupervised detection of structural changes using train – Induced dynamic responses. Mechanical Systems and Signal Processing. 2022;**165**:108268. DOI: 10.1016/j.ymssp.2021.108268

[79] Maes K, Van Meerbeeck L, Reynders EPB, Lombaert G. Validation of vibration-based structural health monitoring on retrofitted railway bridge KW51. Mechanical Systems and Signal Processing. 2022;**165**:108380. DOI: 10.1016/j.ymssp.2021.108380

[80] Zhang G, Tang L, Liu Z, Zhou L, Liu Y, Jiang Z, et al. Enhanced features in principal component analysis with spatial and temporal windows for damage identification. Inverse Problems in Science and Engineering. 2021:1-18. DOI: 10.1080/17415977.2021.1954921

[81] Nie Z, Guo E, Li J, Hao H, Ma H, Jiang H. Bridge condition monitoring using fixed moving principal component analysis. Structural Control and Health Monitoring. 2020;**27**:1-29. DOI: 10.1002/stc.2535

[82] Azim MR, Gül M. Data-driven damage identification technique for steel truss railroad bridges utilizing principal component analysis of strain response. Structure and Infrastructure Engineering. 2021;**17**:1019-1035. DOI: 10.1080/15732479.2020.1785512

[83] Li L, Liu H, Zhou H, Zhang C. Missing data estimation method for time series data in structure health monitoring systems by probability principal component analysis. Advances in Engineering Software. 2020;**149**:102901. DOI: 10.1016/j.advengsoft.2020.102901

Chapter 3

# Perspective Chapter: Simplified Matrix Calculation for Analysis of Girder-Deck Bridge Systems

*Alvaro Gaute-Alonso and David Garcia-Sanchez*

## Abstract

In the design of girder-deck bridge systems, it is necessary to determine the cross-sectional distribution of live loads between the different beams that make up the cross section of the deck. This article introduces a novel method that allows calculating the cross-sectional distribution of live loads on beam decks by applying a matrix formulation that reduces the structural problem to 2 degrees of freedom for each beam: the deflection and the rotation of the deck slab at the center of the beam's span. To demonstrate the proposed method, the procedures are given through three different examples by applying loads to a bridge model. Deflection, bending moment, and shear force of the bridge girders are calculated and discussed through the given examples. The use of the proposed novel method of analysis will result in significant savings in material resources and computing time and contributes in the minimization of total costs, and it contributes in the smart modeling process for girder bridge behavior analysis allowing to feed a bridge digital twin (DT) model based on Inverse Modeling holding the latest updated information provided by distributed sensors. The presented methodology contributes also to speed up real-time decision support system (DSS) demands.

**Keywords:** cross-sectional load distribution, girder bridge decks, optimized matrix method, load distribution factors, structural grillage models

## 1. Introduction

Girder bridge decks are a structural typology commonly used in the design of road and railway bridges, and therefore, any optimization in their calculation has a depth impact on the project phase. As explained in [1], the design process relies solely on the designers' experience, intuition, and ingenuity resulting in a depth cost material, time, and human effort. It is very common to use structural grillage models [2–10] to calculate the cross-sectional distribution of live loads between the different beams that make up the cross section of the deck. Another way to deal with the design and calculation of such decks would be to apply different formulations contained in the bridge design standards that allow approximating the cross-sectional distribution of the bending moment and shear stress caused by live loads through what is known as the load distribution factor "LDF" [11–23]. The "LDF" associated with each case study is conditioned by the type and number of beams, their spacing and length, as well as the existence or not of transverse diaphragms that bring transverse rigidity to the deck.

However, the LDF does not allow determining the distribution of bending stresses on all beams; therefore, the design is oversized. The need for a method to calculate the cross-sectional distribution across all types of beams, without resorting to complex structural grillage models or finite element models in specific structure calculation programs, or to approximate parametric methods based on the "LDF," is one of the authors' motivations for the development of the research work that has given rise to this article.

## 2. Traditional methods for girder bridge deck analysis

Structural grillage models began to be used for the analysis of cross-sectional distribution on beams in the 1960s. These models divide the beam deck into longitudinal and transverse beams (**Figure 1**). Longitudinal beams are responsible for providing the longitudinal bending stiffness of the deck, considering as many longitudinal beams as beams conform to the analyzed beam deck. The structural section of each of the longitudinal beams shall be the result of the section composed of the beam analyzed and the effective depth of the contributing deck with that beam [24, 25]. The cross-sectional distribution of the structural model is provided by the cross-beams and the torsional stiffness of the longitudinal beams. The structural section of the cross-beams corresponds to a rectangular section with an equivalent depth to the slab thickness and a depth according to the discretization used in the grillage models. The analysis of such structural models involves the use of specific structure calculation programs that provide the computing power necessary for the resolution of the proposed matrix problem.

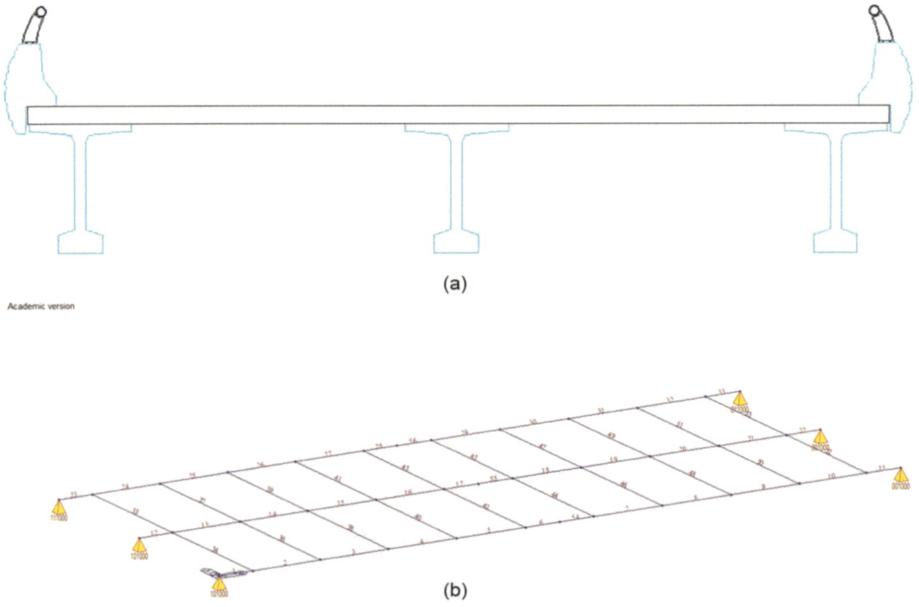


**Figure 1.**
*Grillage discretization: (a) girder bridge deck cross section; (b) structural grillage model.*

The use of structural grillage models allows obtaining the structural response of girder bridge decks to live loads, adequately manifesting the cross-sectional distortion of the girder bridge deck and the distribution of stresses on each of the longitudinal beams that make up the deck. However, such models involve complex, time-consuming analysis that involves the need to use specific structure calculation programs, making it necessary to use simplified methods for the start of the design process.

The concept of LDF was first introduced using empirical formulas at the American Association of State Depthway Officials (AASHO) in 1931 [26]; these methods propose calculating the cross-sectional distribution on beam bridge decks roughly. The LDF is calculated from a series of formulations that parametrically treat the calculation of the percentage of bending moment and shear stress supported by the most requested longitudinal beam. The parameters that condition the calculation of the LDF are the depth of the beam, the span length, the spacing, the number of beams, the position of the load, and the beam position. This method provides an approximate value of the maximum bending stresses on the beams but is not able to reproduce the cross-sectional distribution of longitudinal bending between all the beams that make up the deck.

## 3. Proposed method for the study of the cross-sectional load distribution on a girder bridge deck

The authors propose the use of a method that allows obtaining the cross-sectional distribution of live loads on girder bridge decks without using empirical formulas of LDFs or complex structural models involving the use of specific structure calculation programs. The proposed method is based on using a virtual model that reflects the transverse stiffness of the slab deck, supported on a series of springs that provide the flexural stiffness Eq. (1) and torsional stiffness Eq. (2) of the longitudinal beams that make up the girder bridge deck (**Figure 2**).

$$K_{v,i} = \frac{48 \cdot EI_i}{L^3} \tag{1}$$

$$K_{t,i} = \frac{2 \cdot GJ_i}{L} \tag{2}$$

where $EI_i$ = longitudinal bending stiffness of beam "i"; $GJ_i$ = longitudinal torsional stiffness of beam "i"; L = distance between bridge supports.

The proposed method considers 2 degrees of freedom for each longitudinal beam on the bridge deck: (1) the deflection and (2) the rotation of the deck slab at the center of the beam's span. **Figure 2** represents the structural model scheme for a girder

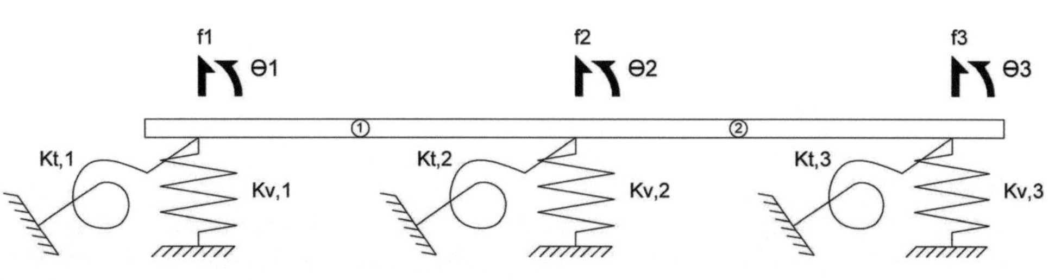


**Figure 2.**
*Proposed method model for cross-sectional distribution on a girder bridge deck.*

bridge deck composed of three longitudinal beams. The matrix approach that solves the structural problem of the cross-sectional distribution of live loads between the different beams that make up the deck is raised in matrix Eqs. (3), (4), (5). The stiffness of the springs that represents the longitudinal bending and the torsional stiffness of the longitudinal beams that make up the model corresponds to the stiffness equivalent to the center of the span.

$$K_e = \frac{EI_e}{L_e^3}\begin{pmatrix} 12 & 6L_e & -12 & 6L_e \\ 6L & 4L_e^{\ 2} & -6L_e & 2L_e^{\ 2} \\ -12 & -6L_e & 12 & -6L_e \\ 6L_e & 2L_e^{\ 2} & -6L_e & 4L_e^{\ 2} \end{pmatrix} = \begin{pmatrix} K_{11,e} & K_{12,e} \\ K_{21,e} & K_{22,e} \end{pmatrix} \tag{3}$$

$$K_{Bi} = \begin{pmatrix} k_{v,i} & 0 \\ 0 & k_{t,i} \end{pmatrix} \tag{4}$$

$$\begin{pmatrix} P_1 \\ M_{t,1} \\ P_2 \\ M_{t,2} \\ P_3 \\ M_{t,3} \end{pmatrix} = \begin{pmatrix} K_{11,1}+K_{B1} & K_{12,1} & \begin{matrix} 0 & 0 \\ 0 & 0 \end{matrix} \\ K_{21,1} & K_{22,1}+K_{11,2}+K_{B2} & K_{12,2} \\ \begin{matrix} 0 & 0 \\ 0 & 0 \end{matrix} & K_{21,2} & K_{22,2}+K_{B3} \end{pmatrix} \times \begin{pmatrix} f_{v,1} \\ \theta_1 \\ f_{v,2} \\ \theta_2 \\ f_{v,3} \\ \theta_3 \end{pmatrix} \tag{5}$$

Where $EI_e$ = transverse bending stiffness of element "e" of the upper deck slab; $L_e$ = length of element "e" of the upper deck slab; $f_{v,i}$ = vertical displacement experienced by longitudinal beam "i"; $\Theta_i$ = transverse rotation of the bridge deck over the longitudinal beam "i."

### 3.1 Loads applied in the center of the span of the longitudinal beams

The maximum longitudinal bending stress in each of the longitudinal beams that make up the bridge deck corresponds to the application of a point load in the center of the beam span. The distribution of the maximum bending moment and maximum shear stress in the different longitudinal beams is obtained by Eqs. (6) and (7), respectively.

$$M_{fm\acute{a}x,n} = \frac{Q \cdot L}{4} \cdot \frac{f_{v,n}}{\sum_{i=1}^{N} f_{v,i}} \tag{6}$$

$$Q_{m\acute{a}x,n} = \frac{Q \cdot L}{2} \cdot \frac{f_{v,n}}{\sum_{i=1}^{N} f_{v,i}} \tag{7}$$

where Q = point load value; L = girder bridge span length.

Example 3.1.1.

In the study of the structural behavior of the bridge deck represented in **Figure 3**, it is desired to know the distribution of the bending moment and the shear force in each of the longitudinal beams generated by the application of the following load states: (a) a point load of 300 KN in the center of the span of beam 1, and (b) a point

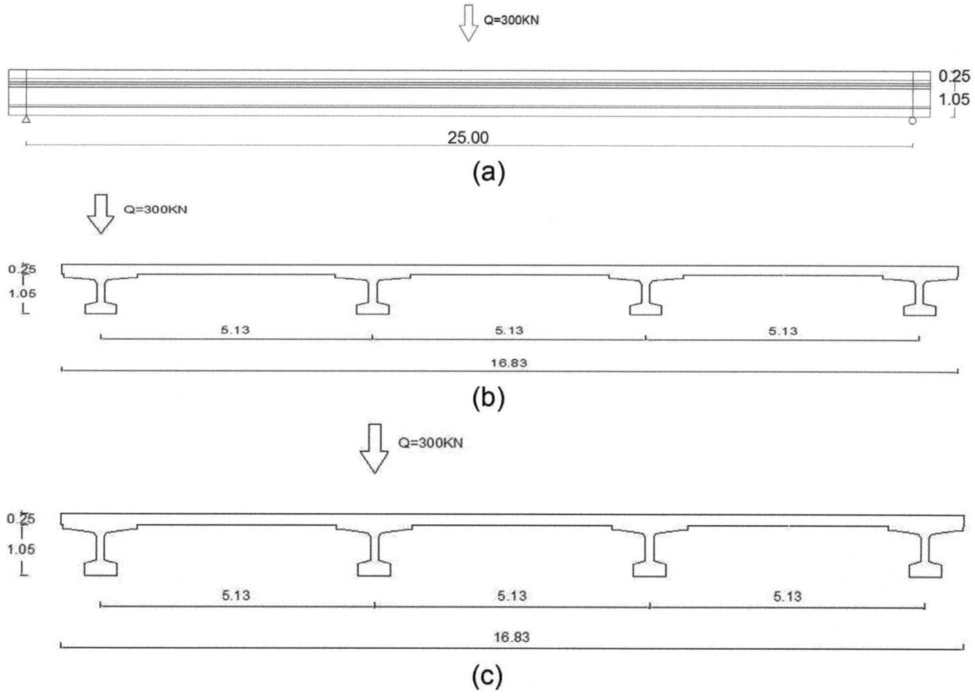


**Figure 3.**
*Girder bridge deck: (a) side view of load states 1 and 2; (b) front view of load state 1; (c) front view of load state 2.*

load of 300 KN in the center of the span of beam 2. The girder bridge is made up of four longitudinal beams 1.05 meters deep (two end beams and two central beams) and an upper slab 0.25 meters thick and 16.83 meters wide. Considering the effective width of the upper slab in each of the longitudinal beams, the inertia to the longitudinal bending of the end and central beams takes the value of 0.1445 $m^4$ and 0.1904 $m^4$, respectively. The longitudinal torsional inertia of the longitudinal beams takes the value of $4.3 \cdot 10^{-3}$ $m^4$. The spacing between longitudinal beams is 5.13 meters, while the distance between the support devices of each longitudinal beam is 25 meters. Both the upper slab and the longitudinal beams are made of structural concrete whose modulus of elasticity reaches 35,000 MPa.

The stiffness of the springs on which the upper slab of the bridge deck rests, and which simulates the bending and torsional stiffness of the longitudinal beams (**Figure 4**), is calculated as follows:

End beams

$$K_{v,end} = \frac{48 \cdot EI_{end}}{L^3} = \frac{48 \cdot 3.5 \cdot 10^7 KN/_{m^2} \cdot 0.1445 m^4}{(25m)^3} = 15,537 KN/_m$$

$$K_{t,end} = \frac{2 \cdot GJ_{end}}{L} = \frac{2 \cdot \frac{3.5 \cdot 10^7 KN/_{m^2}}{2 \cdot (1+0.2)} \cdot 4.3 \cdot 10^3 m^4}{25m} = 5,017\ KN \cdot m$$

$$K_{B,end} = \begin{pmatrix} k_{v,end} & 0 \\ 0 & k_{t,end} \end{pmatrix} = \begin{pmatrix} 15,537\ KN/_m & 0 \\ 0 & 5,017\ KN \cdot m \end{pmatrix}$$

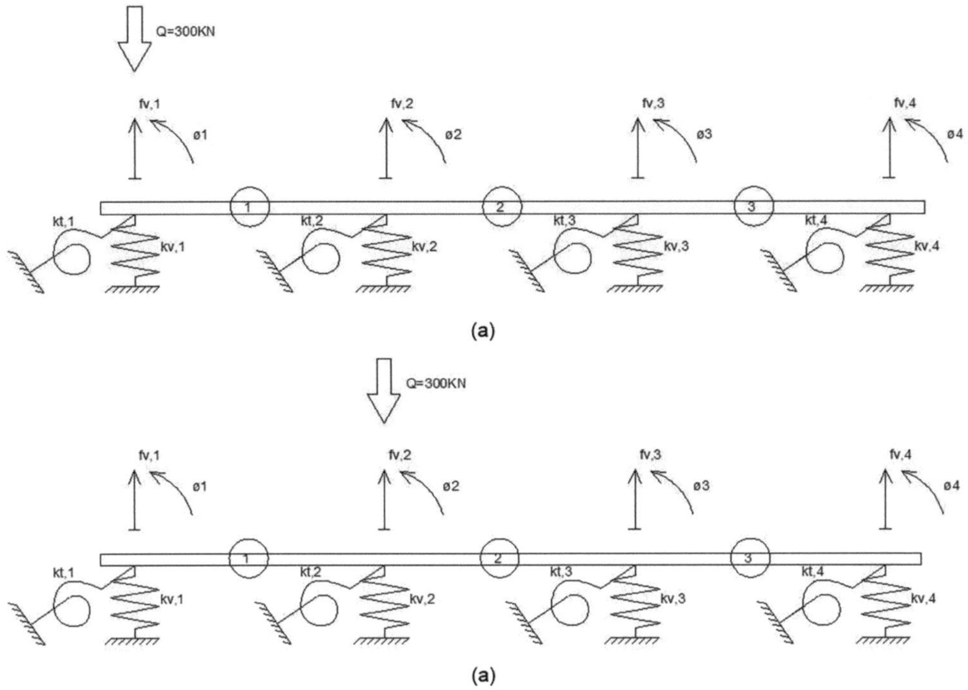


**Figure 4.**
*Proposed structural model: (a) load state 1; (b) load state 2.*

Central beams

$$K_{v,c} = \frac{48 \cdot EI_{end}}{L^3} = \frac{48 \cdot 3.5 \cdot 10^7 \, {}^{KN}\!/\!_{m^2} \cdot 0.1904 m^4}{(25m)^3} = 20,472 \, {}^{KN}\!/\!_{m}$$

$$K_{t,c} = \frac{2 \cdot GJ_{end}}{L} = \frac{2 \cdot \frac{3.5 \cdot 10^7 \, KN/m^2}{2 \cdot (1+0.2)} \cdot 4.3 \cdot 10^3 m^4}{25m} = 5,017 \, KN \cdot m$$

$$K_{B,c} = \begin{pmatrix} k_{v,end} & 0 \\ 0 & k_{t,end} \end{pmatrix} = \begin{pmatrix} 20,472 \, {}^{KN}\!/\!_{m} & 0 \\ 0 & 5,017 \, KN \cdot m \end{pmatrix}$$

The mechanical characteristics of the beam elements that simulate the transverse distribution provided by the upper slab (**Figure 4**) are calculated as follows:

$$I = \frac{1}{12} \cdot 20m \cdot (0.25m)^3 = 2.6 \cdot 10^{-2} m^4$$

$$L = 5,13 \, m$$

$$K = \frac{E \cdot I}{L^3} \begin{pmatrix} 12 & 6L & -12 & 6L \\ 6L & 4L^2 & -6L & 2L^2 \\ -12 & -6L & 12 & -6L \\ 6L & 2L^2 & -6L & 4L^2 \end{pmatrix}$$

$$K = \begin{pmatrix} 81,020 KN/_{m} & 207,800KN & -81,020 KN/_{m} & 207,800KN \\ 207,800KN & 710,690KN \cdot m & -207,800KN & 355,340KN \cdot m \\ -81,020 KN/_{m} & -207,800KN & 81,020 KN/_{m} & -207,800KN \\ 207,800KN & 355,340KN \cdot m & -207,800KN & 710,690KN \cdot m \end{pmatrix}$$

$$K = \begin{pmatrix} K_{11} & K_{12} \\ K_{21} & K_{22} \end{pmatrix}$$

The global stiffness matrix used in the proposed method for the analysis of the transverse distribution of live loads in girder bridge decks is obtained as follows:

$$K_G = \begin{pmatrix} K_{11} + K_{B,end} & K_{12} & \begin{matrix} 0 & 0 \\ 0 & 0 \end{matrix} & \begin{matrix} 0 & 0 \\ 0 & 0 \end{matrix} \\ K_{21} & K_{22} + K_{11} + K_{B,c} & K_{12} & \begin{matrix} 0 & 0 \\ 0 & 0 \end{matrix} \\ \begin{matrix} 0 & 0 \\ 0 & 0 \end{matrix} & K_{21} & K_{22} + K_{11} + K_{B,c} & K_{12} \\ \begin{matrix} 0 & 0 \\ 0 & 0 \end{matrix} & \begin{matrix} 0 & 0 \\ 0 & 0 \end{matrix} & K_{21} & K_{22} + K_{B,end} \end{pmatrix}$$

The transversal distribution of each load state is obtained by planting the compatibility between loads and displacements:

Load state 1

$$\begin{pmatrix} -300KN \\ 0 \\ 0 \\ 0 \\ 0 \\ 0 \\ 0 \\ 0 \end{pmatrix} = \begin{pmatrix} K_{11} + K_{B,end} & K_{12} & \begin{matrix} 0 & 0 \\ 0 & 0 \end{matrix} & \begin{matrix} 0 & 0 \\ 0 & 0 \end{matrix} \\ K_{21} & K_{22} + K_{11} + K_{B,c} & K_{12} & \begin{matrix} 0 & 0 \\ 0 & 0 \end{matrix} \\ \begin{matrix} 0 & 0 \\ 0 & 0 \end{matrix} & K_{21} & K_{22} + K_{11} + K_{B,c} & K_{12} \\ \begin{matrix} 0 & 0 \\ 0 & 0 \end{matrix} & \begin{matrix} 0 & 0 \\ 0 & 0 \end{matrix} & K_{21} & K_{22} + K_{B,end} \end{pmatrix} \times x \begin{pmatrix} f_{v,1} \\ \theta_1 \\ f_{v,2} \\ \theta_2 \\ f_{v,3} \\ \theta_3 \\ f_{v,4} \\ \theta_4 \end{pmatrix}$$

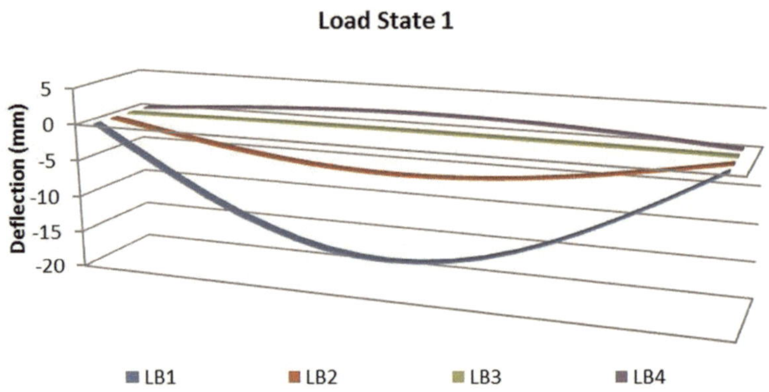


**Figure 5.**
*Deflections of longitudinal beams in load state 1.*

Deflections (**Figure 5**) and rotations

$$d = \begin{pmatrix} -15.14\ mm \\ 2.25\ mRad \\ -4.75\ mm \\ 1.55\ mRad \\ 0.24\ mm \\ 0.52\ mRad \\ 1.79\ mm \\ 0.19\ mRad \end{pmatrix}$$

Maximum bending moments

$$M_{f\acute{m}ax,n} = \frac{Q \cdot L}{4} \cdot \frac{f_{v,n}}{\sum_{i=1}^{N} f_{v,i}} = \begin{pmatrix} 1,588\ KN \cdot m \\ 499\ KN \cdot m \\ -25\ KN \cdot m \\ -187\ KN \cdot m \end{pmatrix}$$

Maximum shear forces

$$Q_{\acute{m}ax,n} = \frac{Q \cdot L}{2} \cdot \frac{f_{v,n}}{\sum_{i=1}^{N} f_{v,i}} = \begin{pmatrix} 127\ KN \\ 40\ KN \\ -2\ KN \\ -15\ KN \end{pmatrix}$$

Load state 2

$$\begin{pmatrix} 0 \\ 0 \\ -300KN \\ 0 \\ 0 \\ 0 \\ 0 \\ 0 \end{pmatrix} = \begin{pmatrix} & & 0 \quad 0 & 0 \quad 0 \\ K_{11}+K_{B,end} & K_{12} & 0 \quad 0 & 0 \quad 0 \\ K_{21} & K_{22}+K_{11}+K_{B,c} & K_{12} & 0 \quad 0 \\ & & & 0 \quad 0 \\ 0 \quad 0 & K_{21} & & \\ 0 \quad 0 & & K_{22}+K_{11}+K_{B,c} & K_{12} \\ 0 \quad 0 & 0 \quad 0 & K_{21} & K_{22}+K_{B,end} \\ 0 \quad 0 & 0 \quad 0 & & \end{pmatrix} \times x \begin{pmatrix} f_{v,1} \\ \theta_1 \\ f_{v,2} \\ \theta_2 \\ f_{v,3} \\ \theta_3 \\ f_{v,4} \\ \theta_4 \end{pmatrix}$$

Deflections (**Figure 6**) and rotations

$$d = \begin{pmatrix} -4.75\,mm \\ -0.76\,mRad \\ -7.21\,mm \\ 0.08\,mRad \\ -4.01\,mm \\ 0.87\,mRad \\ 0.24\,mm \\ 0.80\,mRad \end{pmatrix}$$

Maximum bending moments

$$M_{f\acute{m}ax,n} = \frac{Q \cdot L}{4} \cdot \frac{f_{v,n}}{\sum_{i=1}^{N} f_{v,i}} = \begin{pmatrix} 566\,KN \cdot m \\ 859\,KN \cdot m \\ 478\,KN \cdot m \\ -28\,KN \cdot m \end{pmatrix}$$

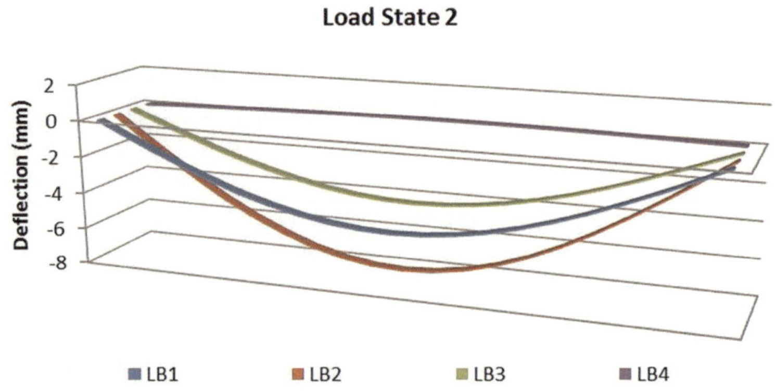


**Figure 6.**
*Deflections of longitudinal beams in load state 2.*

Maximum shear forces

$$Q_{máx,n} = \frac{Q \cdot L}{2} \cdot \frac{f_{v,n}}{\sum_{i=1}^{N} f_{v,i}} = \begin{pmatrix} 45\,KN \\ 69\,KN \\ 38\,KN \\ -2\,KN \end{pmatrix}$$

## 3.2 Loads applied at a distance "x" from one of the supports of the longitudinal beams

The proposed method is applicable to calculate the structural response of a girder bridge deck to the application of a vertical load in any cross section. Using the Maxwell-Betti reciprocity theorem [27], for the application of a point load of value "Q" at a distance "x" from one of the two support points of a longitudinal beam, the deflection at the center of the span is obtained by applying the formulation reflected in Eq. (8). Likewise, the distribution of the maximum bending moment and maximum shear stress in the different longitudinal beams is obtained by Eqs. (9) and (10), respectively.

$$f_{cl,Q(x),i} = f_{v,i} \cdot \sin\left(\frac{\pi \cdot x}{L}\right) \tag{8}$$

$$M_{fmáx,n} = \frac{Q \cdot (L - x) \cdot x}{L} \cdot \frac{f_{v,n}}{\sum_{i=1}^{N} f_{v,i}} \quad / \quad x \leq L/2 \tag{9}$$

$$Q_{máx,n} = \frac{Q \cdot (L - x)}{L} \cdot \frac{f_{v,n}}{\sum_{i=1}^{N} f_{v,i}} \quad / \quad x \leq L/2 \tag{10}$$

Example 3.2.1.

In the study of the structural behavior of the bridge deck analyzed in Example 3.1.1., it is intended to know the deflection in the center of the span of the longitudinal beams, as well as the distribution of the maximum bending moment and the maximum shear force in each of the longitudinal beams generated by the application of the following load states: (a) a point load of 300 KN at a distance equivalent to L/3 from

**Figure 7.**
*Point load of 300 KN at a distance equivalent to L/3 from one of the supports of a beam.*

one of the supports of beam 1, and (b) a point load of 300 KN at a distance equivalent to L/3 from one of the supports of beam 2 (**Figure** 7).

Load state 1

Deflection in the center of the spam of the longitudinal beams

$$f_{cl,Q(x),i} = f_{v,i} \cdot sin\left(\frac{\pi \cdot x}{L}\right) = \begin{pmatrix} -13.12\,mm \\ -4.12\,mm \\ 0.20\,mm \\ 1,55\,mm \end{pmatrix}$$

Maximum bending moments

$$M_{fmáx,n} = \frac{Q \cdot (L - x) \cdot x}{L} \cdot \frac{f_{v,n}}{\sum_{i=1}^{N} f_{v,i}} = \begin{pmatrix} 1,412\,KN \cdot m \\ 443\,KN \cdot m \\ -22\,KN \cdot m \\ -167\,KN \cdot m \end{pmatrix}$$

Maximum shear forces

$$Q_{máx,n} = \frac{Q \cdot (L - x)}{L} \cdot \frac{f_{v,n}}{\sum_{i=1}^{N} f_{v,i}} = \begin{pmatrix} 169\,KN \\ 53\,KN \\ -3\,KN \\ -20\,KN \end{pmatrix}$$

Load state 1

Deflection in the center of the spam of the longitudinal beams

$$f_{cl,Q(x),i} = f_{v,i} \cdot sin\left(\frac{\pi \cdot x}{L}\right) = \begin{pmatrix} -4.12\,mm \\ -6.25\,mm \\ -3.48\,mm \\ 0.20\,mm \end{pmatrix}$$

Maximum bending moments

$$M_{fmáx,n} = \frac{Q \cdot (L - x) \cdot x}{L} \cdot \frac{f_{v,n}}{\sum_{i=1}^{N} f_{v,i}} = \begin{pmatrix} 503\,KN \cdot m \\ 764\,KN \cdot m \\ 425\,KN \cdot m \\ -25\,KN \cdot m \end{pmatrix}$$

Maximum shear forces

$$Q_{máx,n} = \frac{Q \cdot (L - x)}{L} \cdot \frac{f_{v,n}}{\sum_{i=1}^{N} f_{v,i}} = \begin{pmatrix} 60\ KN \\ 92\ KN \\ 51\ KN \\ -3\ KN \end{pmatrix}$$

### 3.3 Loads applied at any point on the bridge deck

The proposed method is applicable to calculate the transverse response of girder bridge decks to the application of a vertical load at any point of the cross section. As in any matrix calculation, if the vertical load acts on a section of slab between longitudinal beams, the degrees of freedom of the structural model are locked and the reactions in the locked degrees of freedom are calculated (rigid step). Subsequently, the degrees of freedom are released and loaded with the reactions obtained in the previous step to obtain the vertical displacement of each of the longitudinal beams that make up the bridge deck (flexible step).

Example 3.3.1.

In the study of the structural behavior of the bridge deck analyzed in Example 3.1.1., it is intended to know the deflection in the center of the span of the longitudinal beams, as well as the distribution of the maximum bending moment and the maximum shear force in each of the longitudinal beams generated by the application of the load states described in **Figure 8**.

Rigid step (**Figure 9**).

The load generated by the action of the truck axles on longitudinal beams 1 and 2 is obtained considering the compatibility between loads and movements in the upper slab between longitudinal beams 1 and 2.

$$\begin{pmatrix} V_{LB1} \\ M_{LB1} \\ -Q_i \\ 0 \\ -Q_i \\ 0 \\ V_{LB2} \\ M_{LB2} \end{pmatrix} = \begin{pmatrix} K_{11,1} & K_{12,1} & \begin{matrix} 0 & 0 \\ 0 & 0 \end{matrix} & \begin{matrix} 0 & 0 \\ 0 & 0 \end{matrix} \\ K_{21,1} & K_{22,1} + K_{11,2} & K_{12,2} & \begin{matrix} 0 & 0 \\ 0 & 0 \end{matrix} \\ \begin{matrix} 0 & 0 \\ 0 & 0 \end{matrix} & K_{21,2} & K_{22,2} + K_{11,3} & K_{12,3} \\ \begin{matrix} 0 & 0 \\ 0 & 0 \end{matrix} & \begin{matrix} 0 & 0 \\ 0 & 0 \end{matrix} & K_{21,3} & K_{22,3} \end{pmatrix} \cdot \begin{pmatrix} 0 \\ 0 \\ v2 \\ \theta 2 \\ v3 \\ \theta 3 \\ 0 \\ 0 \end{pmatrix}$$

Load on longitudinal beams 1 and 2

$$Q = \begin{pmatrix} 1.46 \cdot Q\ KN \\ 1.09 \cdot Q\ KN \cdot m \\ 0.54 \cdot Q\ KN \\ -0.70 \cdot Q\ KN \cdot m \end{pmatrix}$$

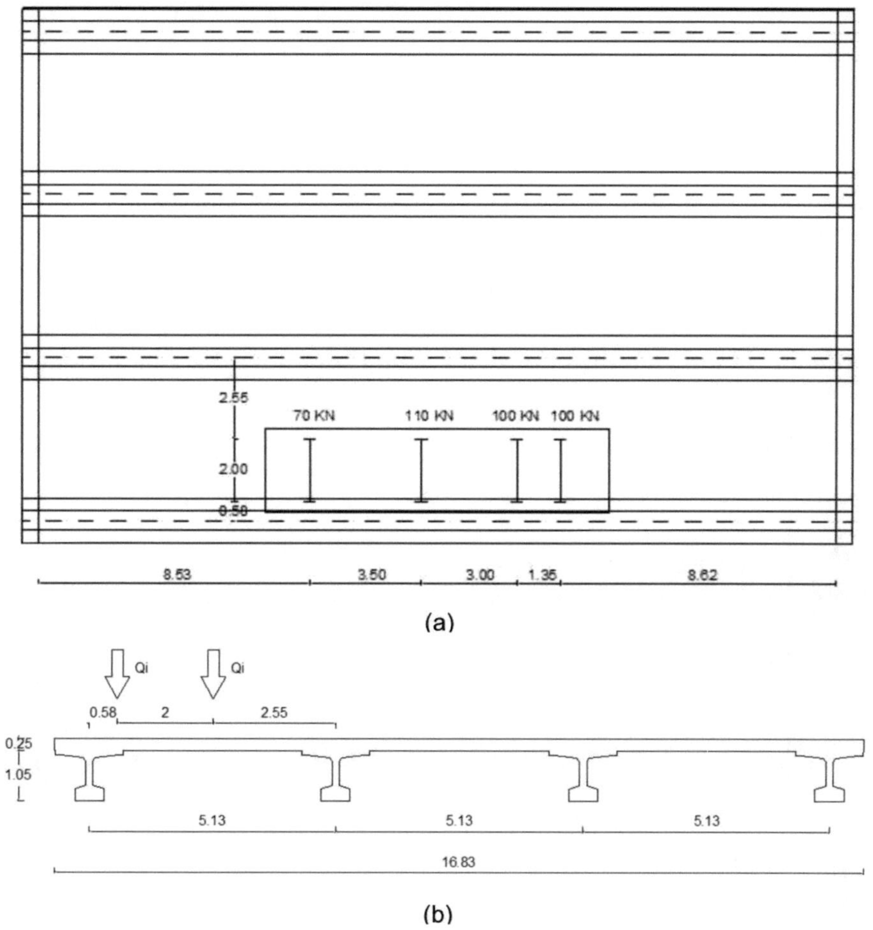


**Figure 8.**
*Load state generated by the actuation of a four-axle truck 70 + 110 + 100 + 100 KN: (a) plan view; (b) front view.*

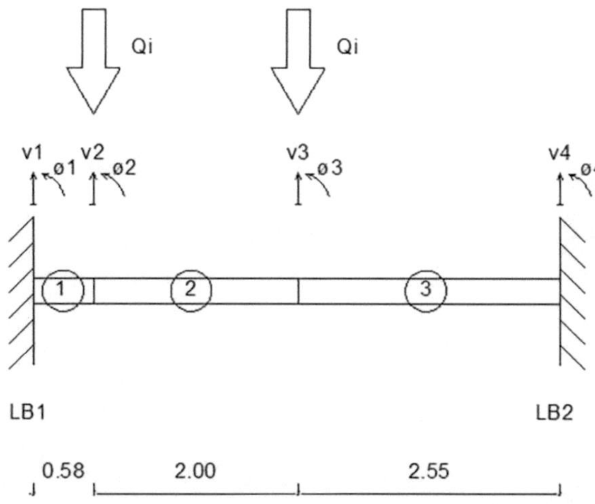


**Figure 9.**
*Load state generated by the actuation of a four-axle truck. Rigid step.*

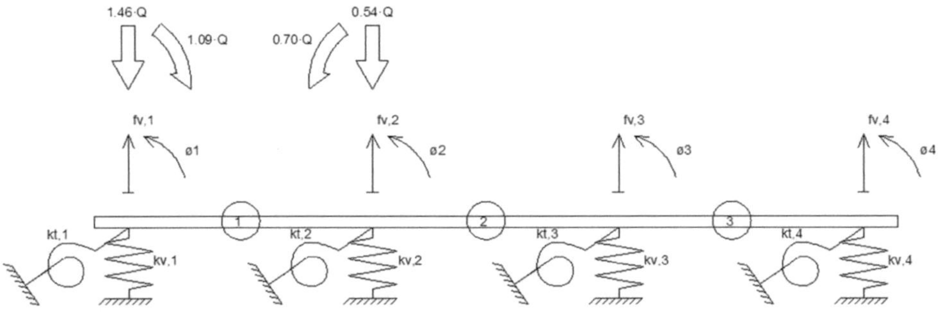


**Figure 10.**
*Load state generated by the actuation of a four-axle truck. Flexible step.*

Flexible step (**Figure 10**).

$$\begin{pmatrix} -1.46Q \\ -1.09Q \\ -0.54Q \\ 0.70Q \\ 0 \\ 0 \\ 0 \\ 0 \end{pmatrix} = \begin{pmatrix} & & 0 \quad 0 & 0 \quad 0 \\ K_{11}+K_{B,end} & K_{12} & 0 \quad 0 & 0 \quad 0 \\ K_{21} & K_{22}+K_{11}+K_{B,c} & K_{12} & 0 \quad 0 \\ & & & 0 \quad 0 \\ 0 \quad 0 & K_{21} & & \\ 0 \quad 0 & & K_{22}+K_{11}+K_{B,c} & K_{12} \\ 0 \quad 0 & 0 \quad 0 & K_{21} & K_{22}+K_{B,end} \\ 0 \quad 0 & 0 \quad 0 & & \end{pmatrix} x \begin{pmatrix} f_{v,1} \\ \theta_1 \\ f_{v,2} \\ \theta_2 \\ f_{v,3} \\ \theta_3 \\ f_{v,4} \\ \theta_4 \end{pmatrix}$$

Truck axle 1
Deflection in the center of the spam of the longitudinal beams

$$f_{cl,Q(x),i} = f_{v,i} \cdot sin\left(\frac{\pi \cdot x}{L}\right) = \begin{pmatrix} -2.39\ mm \\ -1.20\ mm \\ -0.21\ mm \\ 0.30\ mm \end{pmatrix}$$

Maximum bending moments

$$M_{fmáx,n} = \frac{Q \cdot (L-x) \cdot x}{L} \cdot \frac{f_{v,n}}{\sum_{i=1}^{N} f_{v,i}} = \begin{pmatrix} 268\ KN \cdot m \\ 135\ KN \cdot m \\ 24\ KN \cdot m \\ -34\ KN \cdot m \end{pmatrix}$$

Maximum shear forces

$$Q_{máx,n} = \frac{Q \cdot (L-x)}{L} \cdot \frac{f_{v,n}}{\sum_{i=1}^{N} f_{v,i}} = \begin{pmatrix} 31\ KN \\ 16\ KN \\ 3\ KN \\ -4KN \end{pmatrix}$$

Truck axle 2
Deflection in the center of the spam of the longitudinal beams

$$f_{cl,Q(x),i} = f_{v,i} \cdot sin\left(\frac{\pi \cdot x}{L}\right) = \begin{pmatrix} -4.27\ mm \\ -2.15\ mm \\ -0.38\ mm \\ 0.53\ mm \end{pmatrix}$$

Maximum bending moments

$$M_{f\acute{m}ax,n} = \frac{Q \cdot (L - x) \cdot x}{L} \cdot \frac{f_{v,n}}{\sum_{i=1}^{N} f_{v,i}} = \begin{pmatrix} 468\ KN \cdot m \\ 236\ KN \cdot m \\ 41\ KN \cdot m \\ -59\ KN \cdot m \end{pmatrix}$$

Maximum shear forces

$$Q_{\acute{m}ax,n} = \frac{Q \cdot (L - x)}{L} \cdot \frac{f_{v,n}}{\sum_{i=1}^{N} f_{v,i}} = \begin{pmatrix} 39\ KN \\ 20\ KN \\ 3\ KN \\ -5\ KN \end{pmatrix}$$

Truck axle 3
Deflection in the center of the spam of the longitudinal beams

$$f_{cl,Q(x),i} = f_{v,i} \cdot sin\left(\frac{\pi \cdot x}{L}\right) = \begin{pmatrix} -3.69\ mm \\ -1.86\ mm \\ -0.33\ mm \\ 0.46\ mm \end{pmatrix}$$

Maximum bending moments

$$M_{f\acute{m}ax,n} = \frac{Q \cdot (L - x) \cdot x}{L} \cdot \frac{f_{v,n}}{\sum_{i=1}^{N} f_{v,i}} = \begin{pmatrix} 409\ KN \cdot m \\ 206\ KN \cdot m \\ 36\ KN \cdot m \\ -51\ KN \cdot m \end{pmatrix}$$

Maximum shear forces

$$Q_{\acute{m}ax,n} = \frac{Q \cdot (L - x)}{L} \cdot \frac{f_{v,n}}{\sum_{i=1}^{N} f_{v,i}} = \begin{pmatrix} 41\ KN \\ 21\ KN \\ 4\ KN \\ -5\ KN \end{pmatrix}$$

Truck axle 4

Deflection in the center of the spam of the longitudinal beams

$$f_{cl,Q(x),i} = f_{v,i} \cdot sin\left(\frac{\pi \cdot x}{L}\right) = \begin{pmatrix} -3.43\,mm \\ -1.73\,mm \\ -0.30\,mm \\ 0.43\,mm \end{pmatrix}$$

Maximum bending moments

$$M_{fmáx,n} = \frac{Q \cdot (L - x) \cdot x}{L} \cdot \frac{f_{v,n}}{\sum_{i=1}^{N} f_{v,i}} = \begin{pmatrix} 385\,KN \cdot m \\ 194\,KN \cdot m \\ 34\,KN \cdot m \\ -48\,KN \cdot m \end{pmatrix}$$

Maximum shear forces

$$Q_{máx,n} = \frac{Q \cdot (L - x)}{L} \cdot \frac{f_{v,n}}{\sum_{i=1}^{N} f_{v,i}} = \begin{pmatrix} 45\,KN \\ 22\,KN \\ 4\,KN \\ -6\,KN \end{pmatrix}$$

Deflection and efforts generated by the action of the four-axle truck

$$f_{v,i} = \begin{pmatrix} -13.78\,mm \\ -6.94\,mm \\ -1.22\,mm \\ 1.72\,mm \end{pmatrix}$$

$$M_{fmáx,i} = \begin{pmatrix} 1,323\,KN \cdot m \\ 667\,KN \cdot m \\ 117\,KN \cdot m \\ -165\,KN \cdot m \end{pmatrix}$$

$$Q_{máx,i} = \begin{pmatrix} 138\,KN \\ 70\,KN \\ 12\,KN \\ -17\,KN \end{pmatrix}$$

## 4. Conclusions

The proposed method for the analysis of the cross-sectional distribution of live loads on girder bridge decks allows determining the cross-sectional distribution in different configurations of girder bridge decks without the need to resort to complex calculation models that involve depth computing power and excessive analysis time. The method is also applicable for modern synthetic materials, such plastic composites, self-repair. The simplicity of the method allows an easy integration into optimal bridge design strategies [28] or more heuristic approaches [29–33] to challenge today's competitive world intelligently.

## Acknowledgements

This work has received funding from the European's Union Horizon 2020 research and innovation program under the grant agreement No 769373 (FORESEE project). This paper reflects only the author's views. The European Commission and INEA are not responsible for any use that may be made of the information contained therein.

## Author details

Alvaro Gaute-Alonso[1*] and David Garcia-Sanchez[2]

1 Group of Research and Civil Engineering Dynamic Analysis, University of Cantabria "GiaDe", Santander, Spain

2 TECNALIA Basque Research and Technology, Alliance "BRTA", Derio, Spain

*Address all correspondence to: alvaro.gaute@unican.es

## References

[1] Ahsan R, Rana S, Sayeed Nurul G. Cost optimum design of posttensioned I-girder bridge using global optimization algorithm. Journal of Structural Engineering. 2012;**138**(2):273-284. DOI: 10.1061/(ASCE)ST.1943-541X.0000458

[2] Douthe C, Caron J, Baverel O. Gridshell structures in glass fibre reinforced polymers. Construction and Building Materials. 2010;**24**(9): 1580-1589. DOI: 10.1016/j.conbuildmat.2010.02.037

[3] Peloux LD, Tayeb F, Lefevre B, Baverel O, Caron J-F. Formulation of a 4-DoF torsion/bending element for the formfinding of elastic gridshells. Amsterdam: Proceedings of the International Association for Shell and Spatial Structures (IASS) Symposium; 2015

[4] Rombouts J, Lombaert G, Laet LD, Schevenels M. A novel shape optimization approach for strained gridshells: Design and construction of a simply supported gridshell. Engineering Structures. 2019;**192**:166-180. DOI: 10.1016/j.engstruct.2019.04.101

[5] D. Veenendaal and. P. Block. An overview and comparison of structural form finding methods for general networks. International Journal of Solids and Structures. 2012;**49**: 3741-3753. DOI: 10.1016/j.ijsolstr.2012.08.008

[6] Samartin Quiroga A. Notas al cálculo de esfuerzos en tableros de puentes. Hormigón y Acero. 1971;**22**(98):115-135

[7] Manterola J. Cálculo de tableros por el método del emparrillado. Hormigón y Acero. 1977;**28**(122):93-149

[8] Jing-xian S. Application of T-beam Grillage model in reconstruction design of dangerous bridge. 4th International Conference on Mechanical, Control and Computer Engineering (ICMCCE); 2019. DOI: 10.1109/ICMCCE48743.2019.00190

[9] Zhou Y, Ji Y. Comparison and Analysis of the Results of Grillage Method and Single Beam Method to Continuous Box Girder with Variable Width. Banff, Canada: 4th International Conference on Transportation Information and Safety (ICTIS); 2017

[10] Feng JP, Zhang XL, Zhu Z, Huang PM, Wang D, Niu YW. Application on Grillage Method in Box Girder Bridges. Applied Mechanics and Materials. 2013;**1**:886-890. DOI: 10.4028/www.scientific.net/amm.438-439.886

[11] Dicleli M, Erhan S. Live Load Distribution Formulas for Single-Span Prestressed Concrete Integral Abutment Bridge Girders. Journal of Bridge Engineering. 2009;**14**:6. DOI: 10.1061/(ASCE)BE.1943-5592.0000007

[12] Fu CC, Elhelbawey M, Sahin MA, Schelling DR. Lateral distribution factor from bridge field testing. Journal of Structural Engineering. 1996;**122**(9): 1106-1109. DOI: 10.1061/(ASCE)0733-9445(1996)122:9(1106)

[13] Gheitasi A, Harris DK. Overload flexural distribution behavior of composite steel girder bridges. Journal of Bridge Engineering. 2015;**20**(5) 04014076:1-15. DOI: 10.1061/(ASCE)BE.1943-5592.0000671

[14] Hess S, Filosa F, Ross BE, Cousins TE. Live Load Testing of NEXT-D Bridges to Determine

Distribution Factors for Moment. Journal of Performance of Constructed Facilities. 2020;**34**(4) 04020063:1-9. DOI: 10.1061/(ASCE)CF.1943-5509.0001452

[15] J. Huang and. J. Davis. Live load distribution factors for moment in NEXT beam bridges. Journal of Bridge Engineering. 2018;**23**(3). 06017010:1-7. DOI: 10.1061/(ASCE) BE.1943-5592.0001202

[16] Huo XS, Wasserman EP, Zhu P. Simplified method of lateral distribution of live load moment. Journal of Bridge Engineering. 2004;**9**(4):382-390. DOI: 10.1061/(ASCE)1084-0702(2004) 9:4(382)

[17] Idriss RL, Liang Z. In-service shear and moment girder distribution factors in simple-span prestressed concrete girder bridge. Journal of the Transportation Research Board. 2010; **2172**:142-150. DOI: 10.3141/2172-16

[18] Kim YJ, Tanovic R, Wight G. Load configuration and lateral distribution of NATO wheeled military trucks for steel I-girder bridges. Journal of Bridge Engineering. 2010;**15**(6):740-748. DOI: 10.1061/(ASCE)BE.1943-5592.0000113

[19] Semendary AA, Steinberg EP, Walsh KK, Barnard E. Live-load moment-distribution factors for an adjacent precast prestressed concrete box beam bridge with reinforced UHPC shear key connections. Journal of Bridge Engineering. 2017;**22**(11):04017088/1-04017088/18

[20] Yalcin OF, Diclel M. Comparative study on the effect of number of girders on live load distribution in integral abutment and simply supported bridge girders. Advances in Structural Engineering. 2013;**16**(6):1011-1034. DOI: 10.1260/1369-4332.16.6.1011

[21] Harris DK. Assessment of flexural lateral load distribution methodologies for stringer bridges. Engineering Structures. 2010;**32**(11):3443-3451. DOI: 10.1016/j.engstruct.2010.06.008

[22] Kong S, Zhuang L, Tao M, Fan J. Load distribution factor for moment of composite bridges with multi-boxgirders. Engineering Structures.215.110716. 2020;**215** (110716):1-19. DOI: 10.1016/j.engstruct.2020.110716

[23] Terzioglu T, Hueste MBD, Mande JB. Live Load Distribution Factors for Spread Slab Beam Bridges. Journal of Bridge Engineering. 2017;**22**(10) 04017067:1-15. DOI: 10.1061/(ASCE) BE.1943-5592.0001100

[24] G. o. S. Ministry of Public Works. 18.2.1. Ancho eficaz del ala en piezas lineales. EHE - 08. Instrucción de Hormigón Estructural; 2011

[25] G. o. S. Ministry of Public Works. 4.5. Anchura eficaz elástica. RPX-95. Recomendaciones para el proyecto de puentes mixtos para carreteras; 1996

[26] American Association of State Highway and Transportation Officials. Standard specifications for highway bridges. 1st ed. Washington, DC; 1931

[27] Baker WF, Beghini LL, Mazurek A, Carrion J, Beghini A. Maxwell's reciprocal diagrams and discrete Michell frames. Structural and Multidisciplinary Optimization. 2013;**48**(2):267-277. DOI: 10.1007/s00158-013-0910-0

[28] Kuang Y, Ou J. Self-repairing performance of concrete beams strengthened using superelastic SMA wires in combination with adhesives released from hollow fibers. Smart

Materials and Structures. 2008;**17**. DOI: 10.1088/0964-1726/17/2/025020

[29] Ghani SN. Performance of global optimization algorithm EOP for non-linear non-differentiable constrained objective functios. New York: Proceedings of IEEE International Conference on Evolutionary Computation; 1995

[30] Ghani S. A versatile algorithm for optimization of a nonlinear non-differentiable constrained objective function. UKAEA Harwell. R-13714. HMSO Publications Centre; 1989

[31] Hassanain MA, Loov RE. Cost optimization of concrete bridge infrastructure. Canadian Journal of Civil Engineering. 2003;**30**(5):841-843. DOI: 10.1139/L03-045

[32] Jones H. Minimum cost prestressed concrete beam design. Journal of Structural Engineering. 1985;**111**:11. DOI: 10.1061/(ASCE)0733-9445(1985)111:11(2464)

[33] Lounis Z, Cohn MZ. Optimization of precast prestressed concrete bridge beam systems. Precast/Prestressed Concret Institute Journal. 1993;**38**(4):60-78. DOI: 10.15554/pcij.07011993.60.78

Chapter 4

# A Deep Review on a Historical Brick Bridge in South Moravia; Reconstruction and Assessment

*Denisa Boháčová and Eva Burgetová*


## Abstract

The paper describes a structural survey of the brick arched bridge from the 17th century located in Portz Insel near the city Mikulov—in the historic cultural landscape on the Moravian-Austrian border. The bridge is consummate engineering work, originally equipped with a wooden lifting deck, reaching a respectable length of 91.5 m and with 3.7 m. It consists of 15 semi-circular vaulted arches divided by pilasters with double sizes pyramidal edges. It fascinates by its age and it is unparalleled as to the selection of the building material and design. The basic material analysis of the physical and mechanical properties of bridge bricks (moisture, water absorption, resonant frequencies, tensile and compressive strength) were tested including the assessment of the freeze resistance of the original bricks confirming their high quality and durability. The construction was not far from sinking into oblivion until recently when the reconstruction was carried out from 2019 to 2020 and the bridge masonry has been stabilised. Thanks to the inimitable quality of the bricks the construction has survived. The reconstruction of the bridge was awarded the title Monument of the year 2020 by National Heritage Institute.


**Keywords:** historic bridge, structural survey, experimental testing, reconstruction, historic technologies

## 1. Introduction

The large cultural landscape along the border of South Moravia and Lower Austria used to be one continuous area. A varied array of unique monuments related to the evolution of the landscape from ancient times to the present has been preserved. This area became rich thanks to the extraordinary financial potential of the then landowners (the Dietrichsteins, the Liechtensteins, and many other noble families). Their wealth, broad cultural vision, purposefulness, ambition and other contemporary values, along with the effect of positive rivalry, are permanently imprinted on the distinctive character of the landscape.

The project "*Portz Insel Project—Accessing and reconstruction of a designed historic landscape*" was carried out by the City of Mikulov and Drasenhofen in Austria. The preparation of the project began in 2016 and subsequent preparatory steps

**Figure 1.**
*Historic brick bridge.*

were taken. In 2018, a subsidy for the project execution from the cooperation programme *"Interreg V-A Austria-Czech Republic 2014–2020"* was approved. From September 2018 to June 2020 the project itself was carried out. Its main goal was to renovate and provide access to the historic brick bridge with 15 arches and connect the bridge through the renovation of the old road network in the municipalities of Mikulov—Drasenhofen (**Figure 1**).

The completed project aimed to restore the historically valuable area situated in the locality of the peninsula named Portz Insel. It used to be part of the designed peri-urban landscape, whose development was commissioned by Franz Cardinal Dietrichstein in the 17th century. It was one of the earliest generously urbanised landscapes in Moravia and likely also in the larger territory north of the Alps. The island complex, accessible over the long brick bridge supported by arched abutments and connected to the City of Mikulov through a two and a half kilometre long allée, originally an access road for carriages, was an extraordinary work of architecture of its time. It was still described by historians as a landscape jewel of unique beauty in the late nineteenth century [1].

Extraordinary changes spread through the Mikulov region's cultural landscape in the extremely fruitful era of the late Renaissance and early Baroque, associated with the well-known members of the Dietrichstein aristocratic family. Pond cascades, stream and millrace beds, water mills and fisheries, avenues of trees and orchards, bridges and summer houses all created an almost magical landscape of which, regrettably, only fragments have survived. Outstanding among these fragments, the complex near the pond formerly named Portz, harbours the summer residence and ruins amidst the forest. In the 17th century, the complex was surrounded by gardens and cultivated vegetation, offering enjoyment and respite, and showing the status of the Mikulov Princes of Dietrichstein. Franz Cardinal Dietrichstein is a widely known figure of central European history in the early modern times. He would purposefully bring to Moravia the period concepts of Italian Recatholization culture that he was intimately familiar with [1].

## 2. Historic brick bridge

The bridge was located at the narrowest point of the pond (**Figure 2**), still reaching a respectable length of 91.5 m, its width being 3.70 m. It consists of semi-circular

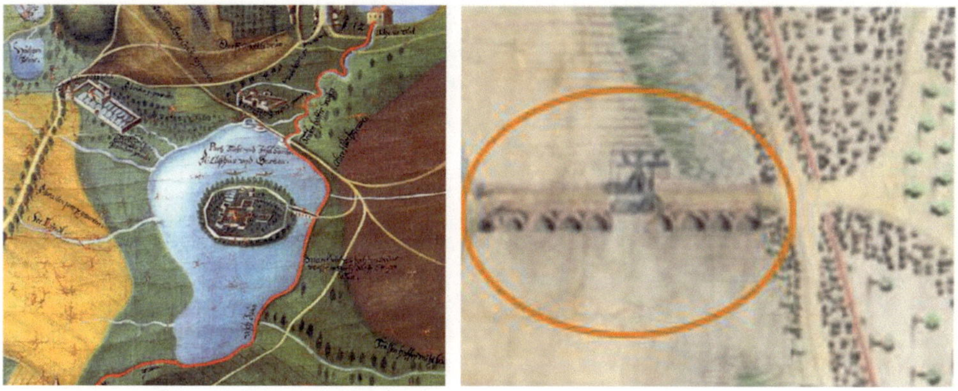

**Figure 2.**
*A view map of the Mikulov estate drawn up in 1672 by Clemens Beuttler of Ebelberg for Prince Ferdinand of Dietrichstein (a map of the Mikulov estate from 1672), the cut out of the Mikulov estate's land register 1629 [1].*

vaulted arches, divided by pilasters with double-sided pyramidal edges (**Figure 3**). The bridge deck was lined with a solid, approx. 75 cm high parapet wall. The outer fully bricked sections were ended by short open wings with decoratively shaped fenders at the ends (**Figure 4**).

The tenth section from the south had originally been equipped with a wooden lifting deck; it was still present on the maps from as late as the 1830s (**Figure 5**), but shortly afterward it was removed and the empty section was bricked up similarly as the other parts of the bridge. After being vaulted, the part of the deck in the originally unfilled segment was also paved with limestone blocks (**Figure 6**) [1].

The existing intact paving in the entire area of the bridge deck made of limestone blocks was found during the restoration, accompanied by an archaeological survey. The paving is well-preserved on about 90% of the surface and during the fieldwork was carefully covered. The stone pavement had an approx. 5–10 cm thick layer of fine limestone gravel on it.

The engineering design of the bridge was very well thought-out. The part of the columns below and slightly above the water level was made of large blocks of hewed limestone, the remainder was built of brick (**Figure 7**). The bridge arches themselves widen at two points towards the base and the bricks on the front face are hewed in the shape of a decorative arch keystone. Also, the pyramidal shape of the edges

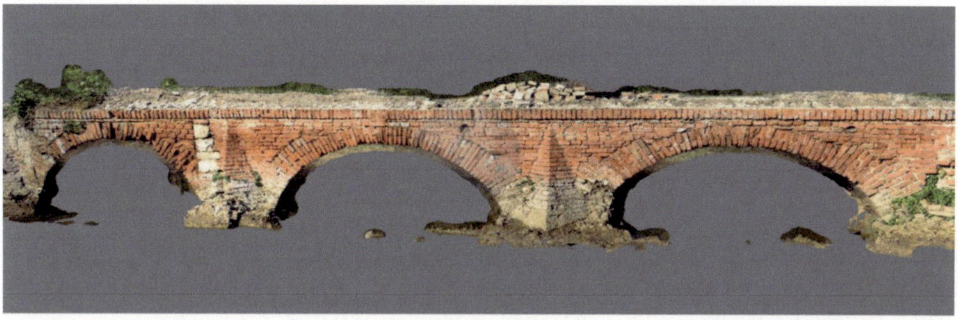

**Figure 3.**
*A cut out of the overall 3D model of the bridge created by laser scanning within an archaeological survey in 2019. A later alteration is visible on the first arch from the left, where the extended vault replaced the old wooden drawbridge [2].*

**Figure 4.**
*The original quoin stone found before the present restoration of the bridge started. The uncovered fender was used in making replicas during the heritage restoration.*

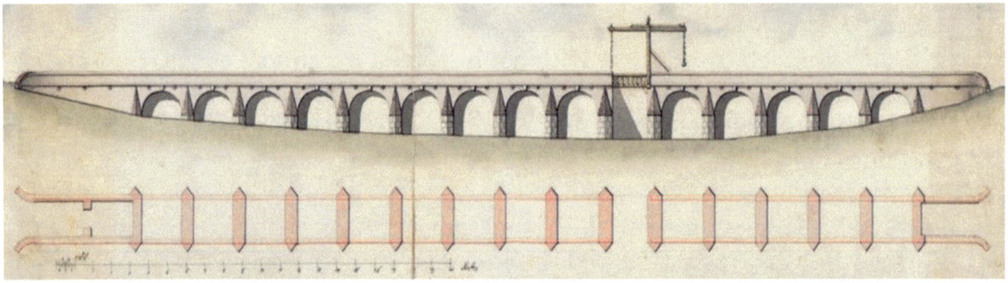

**Figure 5.**
*Historical plan of the original bridge design still with the lifting deck, undated map 107 [1].*

**Figure 6.**
*The find of the original limestone pavement.*

was obtained through the final hewing of the brick masonry to shape (**Figure 8**). According to the documentation, the bridge edges were plastered. On the surface of several bricks, there are still noticeable plaster-coat remains.

Uniquely preserved is the drainage solution for the bridge deck, using drainage openings lined with curved roof tiles, known as tegula. Based on samples of the original historical material of the bridge deck and their laboratory evaluation, it was

**Figure 7.**
*To determine the foundation of the bridge, it was necessary to carry out a borehole probes.*

**Figure 8.**
*A detail of the original edge of the bridge column. A detail of the edge of the bridge column after the repair.*

found that the original bridge deck was threshed on an aluminous-stone backfill with clay waterproofing at the level of the drainage pipes. The thickness of the clay layer was about 60–100 mm.

The Mikulov estate had sufficient quarries supplying stone for construction elements of the bridge as well as for its pavement and penning of the deck. The 17th century land registers record 13 of them. They were in the different localities of the estate, limestone was predominantly quarried between Mikulov and the Portz Island (geological locality Mušlov is a former large sandpit, the main rock type is sands with algal limestone bodies, containing mollusc fauna). Bricks, too, were of local origin, as shown by their marking with the letter N—Nicolsburg (currently Mikulov) (**Figure 9**), a mark or production stamp of identification. The bricks are slightly larger than the current solid fired bricks (approx. 150 mm/300–310 mm).

**Figure 9.**
*Baroque bricks taken out of the loosened structures of the bridge were reused in its restoration. They are marked with the letter N (Nicolsburg).*

Good quality of original bricks was given by the technology of the production process, in the 17th century, building materials were produced more slowly. The bricklayers of that time stuffed the clay into the moulds and let them dry for a month before firing. This production process guarantees high-strength bricks. In the first half of the 17th century there was only one brickyard in Mikulov, but with its three kilns surprisingly large for its time. Given the same technology, it was also possible to burn lime at the brickyards, used in turn to make mortar for the bridge construction. In the mid 17th century, perhaps owing to the consequences of the Thirty Years´ War, only two brick kilns were left in operation.

The golden era of the summer residence complex suddenly came to an end in 1872 when, in consequence of building a railway line to Mikulov (**Figure 10**), the pond

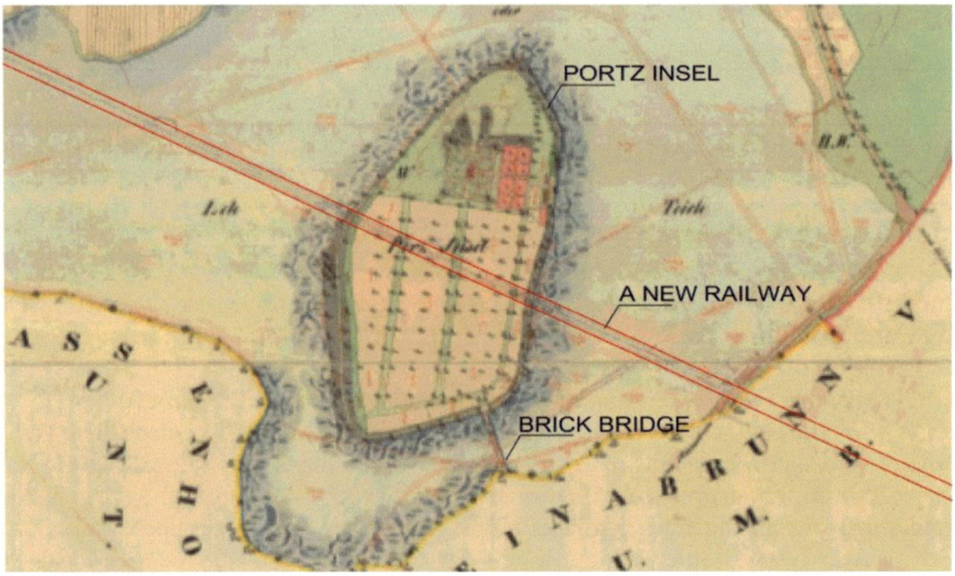


**Figure 10.**
*A new railway line under development (1872) was drawn on the older map. The original pond was then already drained (1855–1857). (Imperial obligatory imprints of stable cadastre map from 1826) [3].*

**Figure 11.**
*A view of the bridge at the time of project preparation before the renovation.*

was drained and the bridge lost its main function. It continued to serve as the crossing over the Rybniční brook. The pond was later restored in the northern surroundings of the island, though due to the railway line it could not be refilled in its original scope, its southern part changed into a wetland forest and meadow. The island became a peninsula and the essentially disused bridge was no longer maintained. Thanks only to the inimitable quality of the bricks made by Mikulov's skilful brick makers of the 17th century, the construction has survived until today. As the whole area had been tightly enclosed over the period of 40 years under the totalitarian regime, the bridge was in a state of disrepair (**Figure 11**), when the restoration was carried out from 2019 to 2020.

The bridge is, in all respects, a unique structure also on a broader European scale. It fascinates by its age, and it is unparalleled as to the selection of the building material and design. It differs from the common pond and river bridges also in its function. Being built as part of a well-considered architectural plan including the other buildings on the island and landscaping, its form elevated the aesthetic effect of the whole majestic complex. The builder had not only in the mind the visual impact of the bridge, but also, through the incorporation of the lifting deck as an element of defensive architecture, he sought to create an impression of a fortress-like character of the place. It is a consummate engineering work that leaves us with the question of who was its originator—perhaps one of the northern Italian fortress master builders, who at that time worked for influential, politically active aristocrats in Bohemia and Moravia.

## 3. Structural survey of the historic brick bridge

One of the fundamental preconditions of the structural survey is detailed analyses of the consequences of the interaction between the external environment and degradation processes in time including and taking into account namely the specificities of repeated cyclic temperature and moisture effects. In prominent cultural and historical monuments, these requirements are of utmost significance and necessity. Protection of structures against these effects is a fundamental part of preservative measures avoiding degradation. Mechanical states of stress due to these effects very often exceeded the stresses and deformation caused by force effects. In the case of moisture, attention must also be paid to chemical and microbiological degradation processes strongly dependent on cyclically changing climatic conditions.

Main objectives of the survey:

- Monitoring and evaluation of non-force effects and influences affect historic structures
- Setting reliability and time dependencies in relation to defined properties of the exterior environment
- Monitoring of effects of the intensity of degradation and durability on applied bricks
- Classification of degradation processes in individual parts of the brick bridge
- Complementation of basic material analysis of the bridge: physical and mechanical properties of bricks (moisture, water absorption, capacity, resistance, tensile and compressive strength, modulus of elasticity)
- Design of efficient direct and indirect rehabilitation measures to be applied in historical bridge structures.

### 3.1 Laboratory testing of bricks

Experimental testing remains the most practised approach towards remedial techniques and information obtained in real buildings can be very useful, especially if the systematic survey was performed.

A total of 14 pieces of original bricks from the 17th century were delivered to the laboratory. The set was marked "N" according to the brand, which most likely means the designation of origin (Nicolsburg—Mikulov). Depending on their appearance and dimensions, the bricks actually correspond to early Baroque bricks—they are lower and wider than usual for bricks from a later period as probably brick N 12 (marked "GN" and other shapes). The bricks, with the exception of brick N 12, had a similar appearance, shard colour and dimensions. Two samples were stored in the archive, 12 pieces of bricks were selected for testing (**Figure 12**).

The samples were divided according to non-destructive tests—ultrasonic and resonant. If the samples were divided into sets only on the basis of a visual impression or even at random, the results could be significantly affected. Based on the long-term experience, the first individual frequency of transverse oscillation was used as crucial for the brick classification. The frequency of oscillation responds very sensitively to both the quality of the material and, in particular, the internal faults in the material, outside invisible.

#### *3.1.1 Brick bulk density test*

The bricks were first dried at 105°C to a constant weight, then they were measured, weighed and the bulk density was calculated in the dried state $\rho_{d,u}$.

$$\rho_{d,u} = \frac{m_{dry,p}}{V_{g,p}} \left[kg \cdot \mathrm{m}^{-3}\right],$$

$m_{dry,p}$ dry sample weight [kg]
$V_{g,p}$ product volume [m$^3$]

**Figure 12.**
*The bricks used for laboratory testing are marked "N" (Nicolsburg). Their appearance and dimensions correspond to early baroque bricks except lower and wider brick N 12 from a later period (marked "GN") [4].*

The bulk density of individual brick samples was relatively uniform, ranging from 1546 to 1641 kg/m$^3$.

### *3.1.2 Brick water absorption test*

Furthermore, they were subjected to a water absorption test, and the water absorption of the bricks was calculated from the difference in weights in both boundary moisture states. The weight absorption was also very balanced—from 15.4 to 18.1% (for a different sample N 12).

### *3.1.3 Brick frequency test*

The bricks were selected for the following sets according to the resonant frequencies as well as the anomalies in the frequency curves:

- Frozen set—bricks N 1, N 3, N 5, N 7, N 10 and N 12,
- Comparative (unfrozen) set—bricks N 2, N 4, N 6, N 8, N 9 and N 11.

In addition to the natural oscillation frequencies, the magnitude of the oscillation amplitude and the sharpness of the frequency curve are also monitored. High oscillation frequency, clear and sharp oscillation amplitude (basically high, clear and clean tone, which the brick sounds when knocking) is typical for quality material without defects, lower frequency together with indistinct or double amplitude, on the contrary, indicates defects or discontinuities in the internal structure of the brick. All samples showed very sharp vibration curves, except for samples N 3, where there were signs of internal structure defects [4].

### *3.1.4 Brick freeze resistance tests for 25 cycles*

According to ČSN 722609 technology of brickmaking, the samples selected for frost resistance tests were first saturated with boiling. Then the resonant oscillation frequencies were measured again. This was followed by alternating freezing and thawing for 25 cycles in an automatic freezing device including the calculation of the so-called relative dynamic modulus of elasticity.

From the results, it is evident that practically all samples passed on frost resistance after 25 cycles, only in sample N 3 found anomalies in the longitudinal resonance frequency. It was a sample that resulted in a greater decrease in resonance frequencies in the saturated state, which can predict a certain disorder in the internal structure. In addition, the samples were dried and subjected to a tensile strength test for bending and strength in pressure.

### *3.1.5 Brick tensile and compressive strength*

The average value of the flexural tensile strength was 3.64 MPa for the comparative samples, and 4.26 MPa for the frozen ones. This is given by the selection criteria into individual sets—the better bricks have been selected for freezing, but it is clear that virtually all samples have safely passed on 25 freezing cycles. The exception is sample N 3 ($f_{b,f}$ = 0.94 MPa), for which resonant test indicated a possible failure in the internal structure. This was indeed reflected in the bending strength test, which was significantly lower than that of the other samples.

According to ČSN 722609, the compressive strength after freezing shall not decrease by more than 15% against the declared compressive strength, in this case against the strength in the pressure of the comparative (non-frozen) bodies. The average value of the bend tensile strength was 22.6 MPa, for frozen bricks up to 24.8 MPa. This is because it is basically unrealistic to create two completely comparable sets of bricks, on the other hand, it is indicative of the excellent frost resistance of the original bricks from the bridge. According to the comparative bodies, the original bricks can be attributed to the strength of the P20 (the mean compressive strength of 22.60 MPa, minimum 19.2 MPa). From the point of view of frost resistance, when the main criterion is the drop of compressive strength, all samples have been passed for 25 cycles [5].

The material analysis confirmed the good quality of the bridge bricks and together with evaluation of the quality of the environment, it provided grounds for establishing the principal degradation agents. During the visual inspection the following failures of the bridge were identified:

- loosened building materials in different parts of the bridge (including a whole line of brickwork in selected section), collapsed parapet walls (**Figure 12**) [4],
- dampness, non-functional waterproofing system (clay sealing and ceramic drains),
- biodeterioration (occurrence of mosses, algae and plants including bushes and trees) (**Figures 11** and **12**).

The historic brick bridge may serve for demonstrating the seriousness of external effects, their growing aggressiveness and the result of mutual structure environment interaction including the surrounding climatic and biological ecosystems and anthropogenous factors [5].

### 3.2 Mechanical failures

Mechanical failures were mainly manifested by cracking ranging from hairlines to prominent tension and shear cracks, going through the joints of brickwork masonry and stone blocks. The extent and intensity of masonry degradation of individual vaults arches varied. Some vaults' arches were damaged by prominent longitudinal cracks up to several millimetres wide which extend through several masonry layers. The respective cracks were mostly situated too close to vault edges. Other cracks mostly local non-continuous ones were situated at various points and did not follow any traceable patterns. Parapet walls were damaged by loose bricks and cracks in the footing bed joint between the vault and the wall.

### 3.3 Moisture analysis of the masonry

In terms of dampness the foundation masonry and pillars, which were currently covered with soil, were the most stressed. The moisture penetrated the structure through direct contact with the soil of adjacent terrain. The high moisture of the masonry was detected at the side of the performed probes. There was direct flooding of the probes. Building materials were disturbed and the binder was being washed off.

Rainwater was the main reason causing the high dampness of the bridge's brickwork. The water from the bridge deck seeped into layers of the backfill up to the vault (the content of moisture reached 15% by weight on top of the vaults). On several places there is no backfill, the rainwater affected directly the vault and further seeped to its lower face. Diversion of water was supposed to be done by the brick drainpipes. Considering the state of the pipes, it was possible that water levels were locally higher within particular sections of the bridge.

One of the possible causes of collapsing of the parapet wall was saturation with water from the entire backfill due to broken and non-functional waterproofing and thus increase of active pressure on the back of the parapet wall and increase of tension on the face side, which masonry was no longer able to transmit.

A very exposed part of the bridge is face surfaces—parapet walls, cornices and masonry railings. The wind-driven rain penetrated the masonry. On the windward side, there is a synergy of negative phenomena, which caused significant degradation of the masonry due to weathering manifested by leaching of the binder and scaling of surface layers. The arches were covered with soil and airborne vegetation.

Based on the analysis the summary of reasons could be stated:

- the intrusion of moisture into the structure,
- random climatic burdens (effects of atmospheric water, wind, cyclic temperature),
- absence in bridge maintenance caused trees rooted in (biodegradation) including anthropogenous factors [5, 6].

## 4. Reconstruction of the historic bridge

Starting summer 2019, a vast reconstruction of the bridge structure had carried out. During the fieldwork after lowering the terrain around the bridge, a problem of much a worse state of the pillars in their lower part, as well as a higher level of groundwater, arose. Some of the pillars had to be statically supported.

The reconstructing the historic bridge is described briefly in the following processes:

Digging up the terrain around the bridge.

Vegetation surrounding the bridge was removed. Subsequently, all historical bricks from the area around the bridge were collected, cleared up and stored. The bricks were cleaned from moss and other vegetation.

Diversion of the stream and remediation of pillars in the original stream bed, damaged bricks replacement.

In order to enable repairs and remediation of the middle pillars of the stream bed, it was crucial to temporarily divert the stream bed. The terrain around the pillars was manually dug away while the vaults were gradually supported (**Figure 14**). Within deeper layers, it had to be needed to drain the leaking groundwater.

Damaged and irregularly shaped bricks had to be taken away carefully and replaced by new replicas produced of at least the same quality compared to the original ones, meaning the quality of "klinker brick these bricks were inserted into cleaned holes after missing or removed ones in order to be statically integrated (activated) at the place of contact with the original brickwork by a special and appropriate expandable mortar mixture—a filler substance intended for caulking of load-bearing joints, the substance is freeze and weather conditions resistant as well." After returning the stream into its original position, gradual remediation of other bridge piers followed.

Repair of fillings above the vaults, remediation of the parapet walls and drainage pipelines (**Figures 13** and **14**).

When dismantling the masonry it was necessary to preserve or avoid breaking the original backfill especially in the area below the level of the drain pipes, meaning under the clay waterproofing. Historical backfill above this level had to be cleared of roots and contaminating soil and preserved for reuse, i.e. to replenish the lower part

**Figure 13.**
*An overall view of the top of the bridge during reconstruction.*

**Figure 14.**
*Dismantling and support of arches, summer 2019.*

of the backfill. Replicas of hand-fired bricks were made after laboratory analysis. A lime mortar was used for masonry. Replicas of the drainpipes were installed in the original places just above the clay waterproofing on the same slope. These replicas were made of fired ceramic precisely according to the preserved remains of the original historical pipes, the exact length was determined on-site so that the face of the parapet walls is exceeded at least 40 mm (**Figure 15**). The pipes were hydrophobized from inside before mounting.

Based on the probes for ascertaining the structure condition below the terrain, it was decided that the soil was removed only to the level of limestone blocks, not to the base of stone packing, in order to avoid possible movements of the stone foundations.

Discovered original bricks were used to complete the lower faces of the vaults. Leaning-out retaining walls at the beginning and the end of the bridge were dismantled and straightened up. The described situation was repeated in other sections—it was essential to perform manual demounting of existing bricks from damaged parts including their cleaning and reusing in particular sections (**Figure 15**). Based on the experimental analysis the original bricks were compared with brick production in a nearby brickyard. Laboratory analysis revealed that the newly produced bricks

**Figure 15.**
*View over the bridge during reconstruction, summer 2019 and detail drainpipes replicas.*

**Figure 16.**
*The bridge after completion, 2020.*

were not suitable for use due to poor frost resistance and it was necessary to find an adequate replacement of another supplier.

Completion of the original fillings and construction of the road threshing surface.

Threshing became the chosen technology for the final surface treatment of the bridge deck, under which is the original threshing layer, stone packing and the backfill [7–9].

The bridge construction that connected the Portz Island to land was not far from sinking into oblivion until recently. Thanks only to the inimitable quality of the bricks made by Mikulov's skilful brick makers of the 17th century, the bridge construction has survived until today (**Figure 16**).

## 5. Conclusions

Based on the presented extensive review on reconstruction and assessment of the historic brick bridge, the following conclusions are drawn:

- Samples of original bricks from the 17th century taken from the "Portz Insel" bridge, have been tested for the assessment of their freeze-resistance. On the basis of the test results, the original bricks from the bridge comply with the requirements for the strength mark P 20. These bricks from the bridge construction have been exposed to external influences over a period of approximately 280 years. At the same time, they are visually very similar, do not differ either of their dimensions nor the volume mass or the absorbency. Practically there are no cracks in them, the brick brittle fracture under tensile test in bending was mostly straight, the ceramic shard is also good quality and well-burned. It is obvious that these are extremely well-produced and durable bricks.

- Currently, the bridge masonry has been stabilised. The accuracy of the results critically depends on the availability of all input parameters (materials features, meteorological data and soil conditions), hence, the results could be used to predict the behaviour of existing structures. Nevertheless, experimental testing remains the most practised approach, especially if the systematic survey was performed.

- The main benefit of the project is the suggestion in using historical technologies and traditional material solutions for the purpose of reconstruction (bricks, clay sealing in combination with ceramic drainage). The proposed historical materials are inert, environmentally friendly and renewable with long service life. The project shows that historical materials together with modern technologies can be profitable for optimal realisation the consummate bridge was saved for the future.

The bridge restoration was awarded for the best reconstruction of the year 2020 by the National Heritage Institute.

## Acknowledgements

The results presented in this article were obtained in the framework of institutional research of the Faculty of Civil Engineering CTU in Prague. Stated facts were analysed in relation to the design and realisation of the reconstruction of the historic bridge located in Portz Insel. Acknowledgements belong to everyone who participates in the reconstruction with the financial support of the Project Interreg Austria—Czech Republic.

## Author details

Denisa Boháčová* and Eva Burgetová
Faculty of Civil Engineering, CTU in Prague, Prague, Czech Republic

*Address all correspondence to: dbfirast@gmail.com

## References

[1] Collection F18. The Main Filling Cabinets of the Dietrichsteins in Mikulov

[2] Tejkal M. Documentation of the Building Monument's State of Discovery by Ground-Based Laser Scanning and Photogrammetry at the Beginning of the Reconstruction. Dolní Kounice; 2019

[3] Central Surveying Archives and Cadastre of Moravia and Silesia, Brno

[4] Test report on bricks in the event Mikulov—Portz Insel, ordered Plus, s.r.o. 2019

[5] Libecajtová A. Numerical analysis of compressed masonry columns. Periodica Polytechnica Civil Engineering. 2020;**64**(3):722-730. ISSN: 0553-6626

[6] Wasserbauer R, Rácová Z, Loušová I. The effect of algal and bacterial colonies on the formation of corrosion active compounds degrading silicate. Building Materials. Chemické listy. 2015;**109**:718-721. ISSN 0009-2770

[7] Boháčová D, Boháč R. Structural Survey and Project Documentation. Prague/Mikulov—Portz Insel; 2016-2019

[8] Korandová K, Hromek J. Municipal Information Office Mikulov. 2016-2017

[9] Biely B, Hromek J. Notes From Site Controls. Mikulov—Portz Insel; 2019-2020

**Chapter 5**

# Challenges in the Construction of Highways in the Brazilian Amazonia Environment: Part I – Identification of Engineering Problems

*Nilton de Souza Campelo,*
*Arlene Maria Lamêgo da Silva Campos,*
*Marcos Valério Mendonça Baia, Daniel Jardim Almeida,*
*Raimundo Humberto Cavalcante Lima,*
*Danielly Kelly dos Reis Dias,*
*Júlio Augusto de Alencar Júnior and*
*Mário Jorge Gonçalves Santoro Filho*

**Abstract**

The construction of highways in the Brazilian Amazonia Region is always problematic, mainly because it involves environmental obstacles but also technical, economic, and natural challenges. The environmental issues concern the deforestation of the virgin forest and the resulting environmental impacts. The technical ones are related to the natural subgrade, formed by the geologically young alluvial soils that are plastic, being highly compressible or expansive, present in the vast Amazon Basin, whereas the economic issues refer to the final costs of inputs for the construction of the layers of the highway since granular soils and stony materials are located in limited areas that are distant from the work sites, given the geographic immensity of the Brazilian Amazonia. There is also the cost of purging low-bearing capacity soil from the natural subgrade of the highway. Added to all this are the issues of nature, which involve high annual rainfall and the hydrological regime of river flooding and ebbing, which induce the saturation of the pavement layers and the loss of the global geotechnical stability of the compacted earth embankment, respectively. This work points out the Engineering difficulties to be faced in road infrastructure works in the Brazilian Amazon.

**Keywords:** highway, pavement, soft soil, expansive soil, synthetic coarse aggregate of calcined clay (SCACC), reinforced piled embankment, recycled material, Brazilian Amazonia

## 1. Introduction

Brazil is a large country of 8.5 million km$^2$ [1]. The International Amazonia is vast, covering parts of nine countries (Brazil, Bolivia, Colombia, Ecuador, Guyana, French Guiana, Suriname, Peru, and Venezuela), equivalent to 8 million square kilometers of South America (**Figure 1**), of which approximately 65% is located in Brazil [2, 3]. The Brazilian population in this region is approximately 20 million [1].

Brazil has a wide climatic and geomorphological variety. This variety is responsible for the presence of several important biomes and ecosystems, which are home to approximately 20% of the living species known worldwide. It is estimated that there are approximately 2 million species of plants, animals, and microorganisms in Brazil [4]. The most important biomes of Brazil are the Amazon Forest and deciduous forests in the North, the rainforest of the Eastern Coast (known as the Atlantic Forest), the savanna areas (*Cerrado*) in the Centre, the thorn forest (*Caatinga*) in the Northeast, the *Pantanal* in the Midwest, and the pine forests and *Pampa* fields in the South. Also noteworthy are the humid riparian forest in the Northwest (*Campinarana*), coastal mangroves, sand dunes, and salt marshes, all transition zones, and many small areas where special combinations of climate, altitude, and soil produce unique ecosystems [4, 3]. **Figure 1** displays these main biomes.

Brazil is at the top of 18 megadiverse countries, home to 15–20% of the world's biological diversity, with more than 120,000 invertebrate species, approximately 9000 vertebrates, and more than 4000 plant species [5]. The Brazilian flora comprises approximately 55,000 described species [6, 7], a number that represents

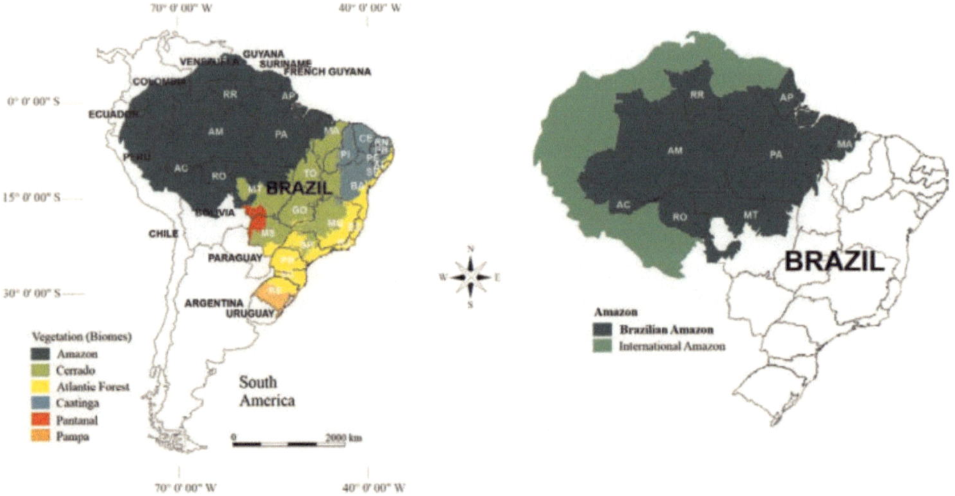


**Figure 1.**
*Distribution of the main biomes of the International Amazonia in South America [3].*

approximately 22% of the world's total species [8]. The Brazilian fauna is also very diverse, with approximately 524 species of mammals, 517 of amphibians, 1622 of birds, 468 of reptiles, more than 3000 species of freshwater fish, and 10–15 million species of insects [8].

The climate of the International Amazonia is classified as humid equatorial—Af, Am, and Aw [9], as shown in **Figure 2a**. The average temperature is between 24 and 26°C, and the annual range is between 1 and 2°C. The rainy season of most of the Amazon Basin is between November and May, and the dry season runs from June to September. In the rainiest months, the relative humidity varies between 80 and 90%, and during the dry season, it reaches at least 75% [10]. **Figure 2b** shows the average annual rainfall accumulation throughout the region. Maximum precipitation values above 2500 mm per year are observed in the Northwest sector of the Amazon Basin and on the North coast [11].

The Amazon River Basin (ARB) hosts the largest tropical forest and natural drainage basin on the planet (formed by large rivers and those of smaller volume, locally known as *igarapés*). Its drainage area covers more than a third of the South American continent, and the discharge of the Amazon River is responsible for almost a fifth of the total discharge of all rivers in the world. This basin contains several subbasins, the most important being the Negro and Madeira Rivers (**Figure 3**) in the North and Southwest, respectively [12]. As seen in this figure, the Amazon River Basin is a large plain, with a predominant altitude below 250 m. The extents of the areas periodically flooded in these regions depend on the size of the basins, the river flow, the slope, and the geomorphology of the adjacent lowlands, and they vary in width from a few hundred meters to 100 km [13]. The vertical amplitudes of the flood pulses [14] are larger in the central part of the Amazon Basin, reaching approximately 10 m, near the confluence of the Solimões and Negro Rivers ("Meeting of the Waters"), but may reach up to 15 m in some regions [15]. These floods affect many roads in the region, lowering the stability of the road embankments, especially in the ebb cycle (lowering of the water level of the watercourse, lasting 4–6 months), which may induce the

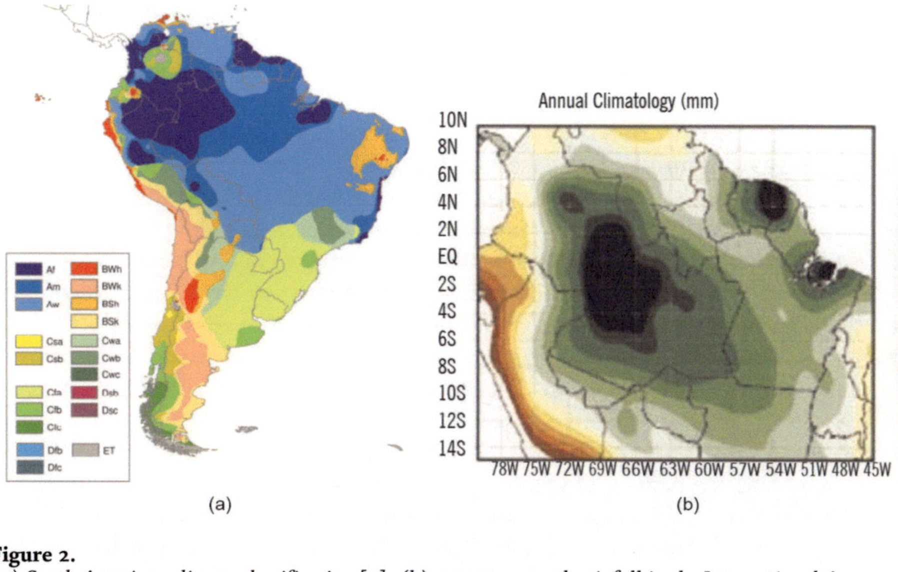


**Figure 2.**
*(a) South American climate classification [9]; (b) average annual rainfall in the International Amazonia [11].*

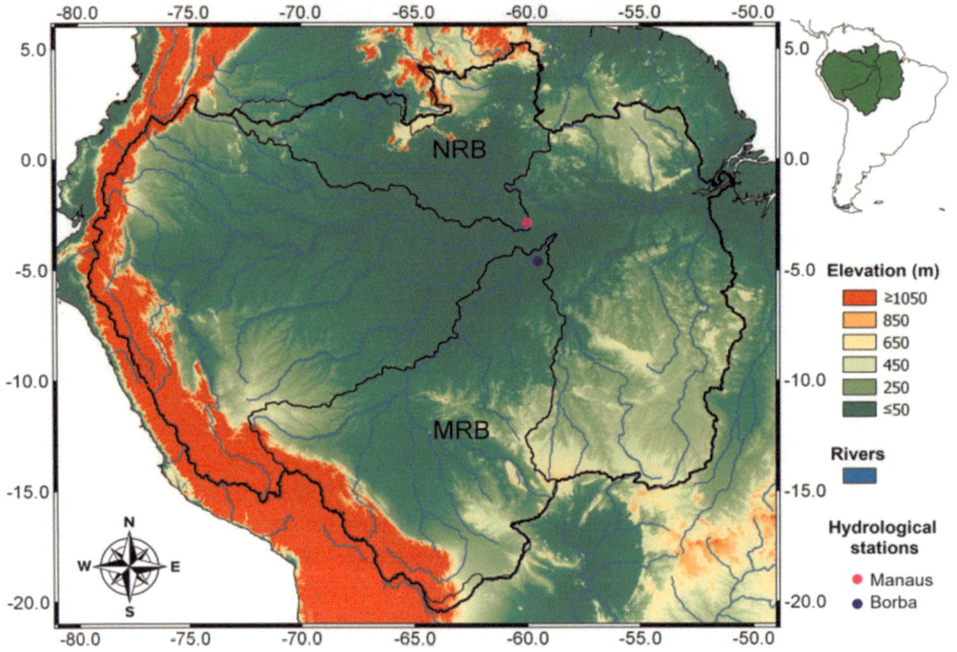


**Figure 3.**
*Amazon River basin (ARB) and its respective elevations [12]. Negro River Basin (NRB) and Madeira River Basin (MRB).*

rupture of the natural slopes and embankments [16]. Approximately 72% of the area of the Amazon Basin has a slope between 0 and 8% [17].

Upland forests represent approximately 83% of the Amazon Basin and are located above the maximum levels of the seasonal flooding of rivers, lakes, and large streams. The floodplain forests are seasonally flooded by nutrient-rich white-water rivers for 6–8 months, and water level fluctuations can reach up to 15 m, covering approximately 7% of the Amazon Basin [18]. The floodplain areas have different altitudes in the interior of the basin, all lower than 30 m altitude, close to the "Meeting of the Waters." **Figure 4** shows a typical cross section of a river in the Amazon Basin [19].

In the Amazonia, approximately 75% of the topsoil (0–50 cm) of the region is formed predominantly by a fine fraction of silts and clays [17], in both the upland (*terras firmes*, **Figure 4**) and lowland areas (*várzeas*). The latter are recent alluvial deposits, with a high amount of organic matter incorporated in the voids; sometimes, these deposits of low bearing capacity extend for tens of meters below the ground surface. Often, mineralogy shows the presence of expansive clay minerals, which show great shrinkage after drying (**Figure 5a**); sometimes they have a dispersed structure [20] that liquefies under the vibration of earthmoving equipment (**Figure 5b**). Generally, *terra firme* soils show pedogenetic evolution of lateralization (colors in red tones) (**Figure 6a**), but this process is not observed for *várzea* soils (**Figure 6b**).

Much of the soil diversity in the Amazon originated from the considerable differences in geology and the history of geomorphology that have arisen throughout the ARB [21]. These authors classified soils according to the main factors that condition their morphological, chemical, and physical properties. Thus, to demonstrate the diversity of Amazonian soils, each of the 14 different reference soil groups surveyed

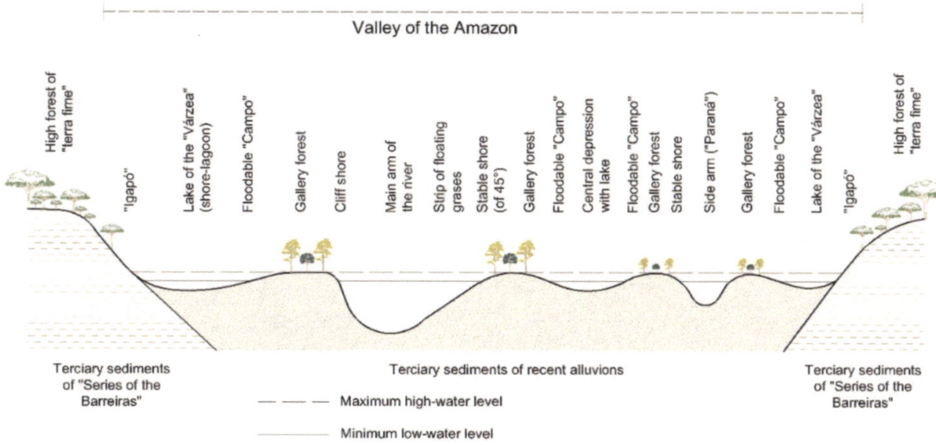


**Figure 4.**
*Schematic representation of some local denominations of flooded and non-flooded areas in the Amazon Basin (adapted from Ref. [19]).*

**Figure 5.**
*(a) Topsoil after natural drying. (b) Loss of resistance of soil with a dispersed structure (right) (photos: NS Campelo).*

was summarized by limited weathering age, humid tropical climate, topography, and drainage and source material.

The natural drainage network formed by the ARB makes the rivers the true "roads" that connect the various interior cities to the capitals. In the Legal Amazon (formed by nine Brazilian States), the few existing federal roads (**Figure 7**) were opened slightly more than 60 years ago as part of the physical integration project of the national territory [22], in a time of strong deforestation for agrarian colonization, opening of fronts for agriculture, and predatory logging, a time that was characterized by the lack of concern about the deforestation of the native forest.

Deposits of commercial rocky materials are limited to regions outside the Amazon Sedimentary Basin. Therefore, the coarse aggregate (crushed stone or pebble) is an input that makes paving services more expensive, given the large distances it needs to cover [23, 24]. **Figure 8** shows the locations of the pebble extraction and crushed stone exploration areas in the state of Amazonas.

(a) (b)

**Figure 6.**
*Vicinal roads were built on (a)* terra firme *(upland area) (lateritic soil), and (b) on* várzea *(floodplain area) (photos: NS Campelo).*

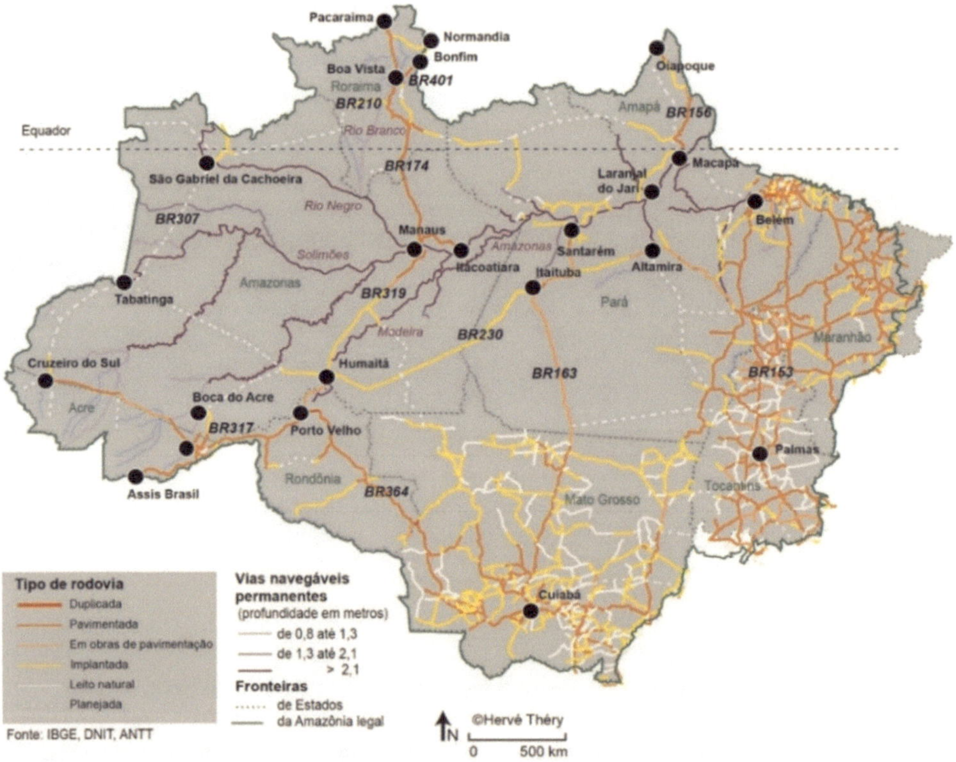


**Figure 7.**
*The most important federal and state highways are located in the Brazilian Legal Amazon [22].*

In the Brazilian Amazonia, roads increase access to the forest, which is followed by deforestation with its ecological impacts [25–27]. The main roads have opened forest areas for settlement and resource extraction [28] and agricultural and timber activities [4], and most of the deforestation occurs in the areas less than 100 km from the main highways under the federal development program, which concentrates almost 90% of

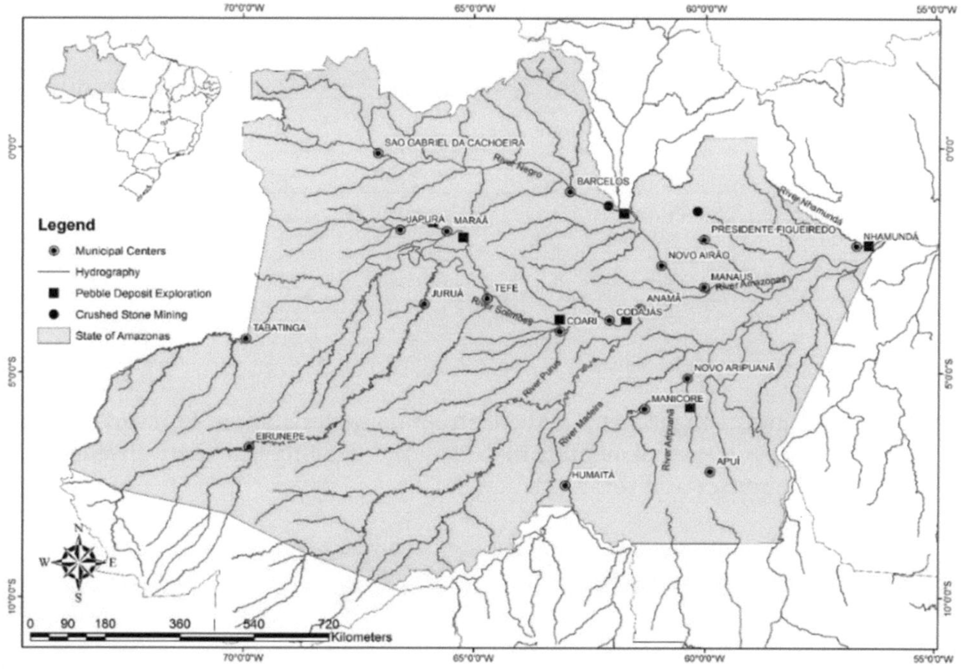


**Figure 8.**
*Locations of natural coarse aggregate (crushed rock and pebble) extraction in the State of Amazonas, Brazil [23].*

the deforestation measured [29]. However, there are more serious cases, in which 94.9% of all deforestation analyzed occurred in a well-defined accessible zone within 5.5 km of some type of road or 1.0 km of a navigable river [28].

Brazil has approximately 1.7 million km of federal, state, and municipal highways, of which only 12.4% are paved [30]. In the Legal Amazon, there are approximately 274,000 km of highways (**Figure** 7), and if the same proportion applies as does at the national level, then only 34,000 km of them are paved.

The National Department of Transportation Infrastructure (DNIT) is the official regulator of Brazilian federal highways. According to the requirements of this organ, the granular layers of the pavement should have a minimum thickness of 15 cm, with minimum California Bearing Ratio (CBR) values and maximum expansions provided in **Table 1**, as a function of the pavement layer.

Thus, as seen in the previous paragraphs, the construction of highways in Brazilian Amazonia is problematic. The obstacles can be grouped into:

- Natural: high annual rainfall; large natural drainage basin (rivers, lakes, channels, and *igarapés*), with large floodplains (*várzeas*) with annual flood cycles of 6 months or more; high vertical amplitude of flooding, approximately 10–15 m; low slope of the region (<8%); lack of natural stone material; silty and clayey topsoils, with low permeability, sometimes associated with expansive clay minerals with a dispersed structure;

- Technical: natural foundation ground (subgrade) composed of soils with low bearing capacity and poor drainage;

| Layer | CBR | Compaction energy | Expansion | Standard |
|---|---|---|---|---|
| Base | ≥80%, for N > 5·106 | Modified | ≤0.5% | DNIT 141/2010—ES |
| | ≥60%, for N ≤ 5·106 | | | |
| Subbase | ISC ≥ 20% | Intermediate | ≤1% | DNIT 139/2010—ES |
| Reinforcement of the Subgrade | CBR > than that of the Subgrade | Normal | ≤1% | DNIT 138/2010—ES |
| Subgrade | CBR ≥ 2% | Normal | ≤2% | DNIT 108/2009—ES |

**Table 1.**
*CBR and pavement layer expansion values.*

- Economic: inputs (aggregates and construction materials) are expensive due to the lack of occurrence of suitable materials and the long transport distance between the source and the construction site;

- Environmental: deforestation and ecological impacts on fauna, flora, soil, and water quality.

## 2. Literature review

### 2.1 Regional and local geology, pedology, and natural drainage of the Brazilian Amazonia

The Amazon Sedimentary Basin (ASB) is an intracratonic sedimentary unit that borders two main areas of the Archean-Proterozoic basement—to the North, the Guianas Craton, and to the South, the Brazil-Central Craton [31]. The ASB is geologically characterized by an extensive Phanerozoic sedimentary cover distributed in the Acre, Solimões, Amazonas, and Alto Tapajós Basins, which were deposited on a Precambrian rocky substrate, where rocks of igneous, metamorphic, and sedimentary nature predominate [32]. **Figure 9** shows the tectonic map of South America [33].

With a drainage area of $6.10^6$ $km^2$, the Amazon River Basin (ARB) is the largest hydrographic basin in the world, covering approximately 5% of the planet's land. The Amazon River has an average annual flow of approximately $210.10^3$ $m^3/s$, contributing approximately 20% of the annual global freshwater discharge to the ocean. Considering its enormous scale, it is not surprising that among the 10 largest rivers in terms of water discharge in the world, four mega-rivers (defined as those with a mean annual discharge $>17{\cdot}10^3$ $m^3/s$) flow into the ARB (i.e., the Amazon, Madeira, Negro, and Japurá rivers), and 24 of the 34 largest tropical rivers also flow through it [34].

The Amazon River rises in the Eastern Cordillera of the Peruvian Andes, at an altitude of approximately 5300 m, and throughout its course, it has many tributaries, the most important of which are the Ucayali and Napo in Peru, and the Javari, Juruá, Purus, Madeira, Tapajós, Xingu, Içá, Japurá, Negro, Trombetas, and Jari in Brazil. In part of the interior of Brazil, the Amazon River is called the Solimões and has, as tributaries of the left bank, the Putumayo-Içá and Caqueta-Japurá Rivers that were born in the Andes of Colombia. On the right bank is the Javari River, which marks the border between Brazil and Peru; the Jutaí, located in Brazil; and the Juruá and Purus, with their sources in Peru. Near the city of Manaus, State of Amazon, the Solimões

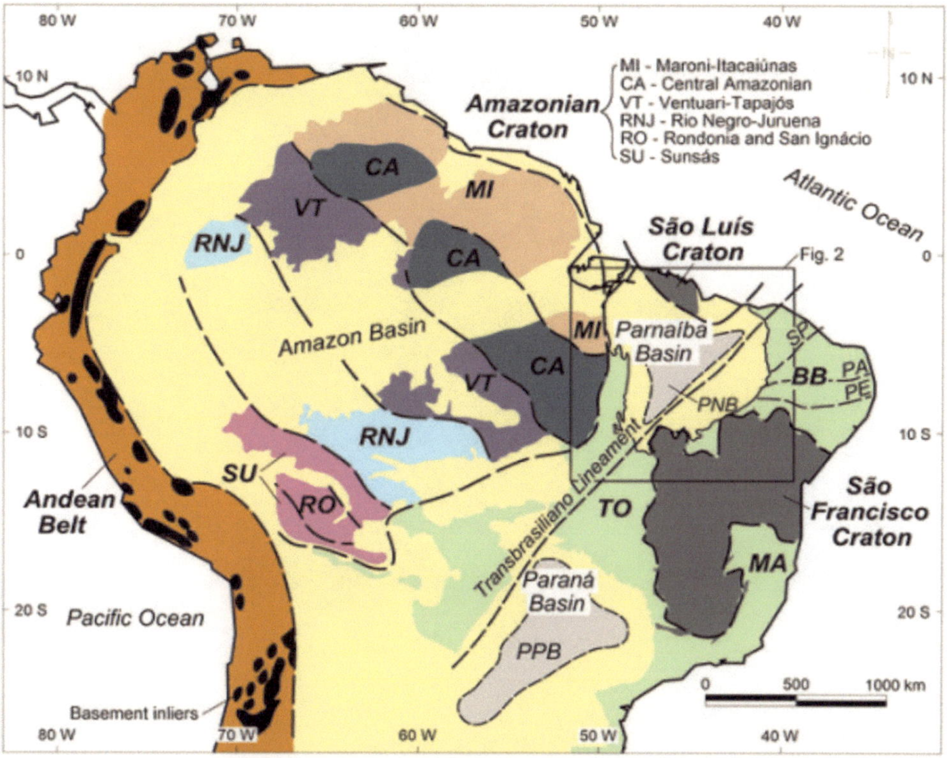


**Figure 9.**
*Simplified tectonic map of northern and southern South America [33].*

River, together with the Negro River, forms the Amazon River, in what was conventionally called the "Meeting of the Waters." The Negro River rises in Colombia at an altitude of approximately 1660 m. The Madeira River, which drains the Eastern Andes of Bolivia and Peru downstream of Manaus, joins the Amazon River on its right bank [35–37]. **Figures 7** and **8** show part of the tributaries of the ARB.

This region is characterized by a great diversity of aquatic environments gathered in the same watershed. The variety of environments is related to the size of the natural drainage area and their strong relationship with environmental factors, relief, pedology, soil, climate, and the different types of vegetation present around the rivers and streams, which are responsible for the notable difference in the composition physics and chemistry of waters [38–41].

In the Shield region, the Amazonian soils are well drained, but in the ASB, in relation to their drainage capacity, they may appear poorly drained, imperfectly drained, or well drained [42, 43]. **Figure 10** shows the main groups of soils found in the International Amazonia [21]. The main classes of soils found in the Brazilian Amazonia are latosols (oxisols) and argisols (ultisols), representing approximately 75% of the superficial soils of the region [44]. Schaefer et al. [45] stated that the distribution of Amazonian soils is marked by geomorphological control—upland and flattened residual geoforms of low plateaus are commonly associated with red–yellow latosols in areas of crystalline rocks or with yellow latosols in areas of Tertiary sediments. In the middle and lower thirds of the hills or flattened residuals, there are argisols, with or without plinthite or petroplinthite, as well as quartzarenic neosols

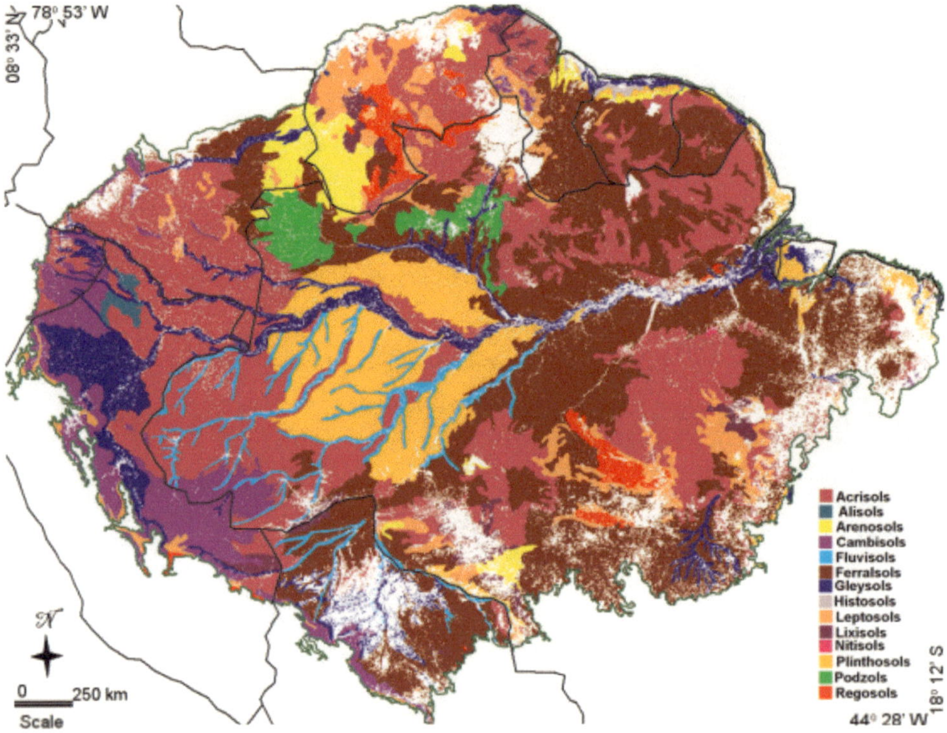


**Figure 10.**
*Soil distribution map of the Amazon Basin, based on the SOTERLAC-ISRIC database [21].*

and spodosols. In the floodplain of white-water rivers, gleysols and Fluvic neosols predominate. Plinthosol soils predominate in the lowlands of the Upper Amazon River and in the Madeira/Purus/Juruá and Solimões/Japurá interfluves.

The chemical and mineralogical characteristics of Amazonian soils are largely dictated by the nature of the source material. Extensive areas of rich and eutrophic soils only exist where there is a current (alluvial plain) or past influence (terraces and low plateaus of the Acre and Upper Amazon River Basins) of Andean sediments or where rocks of higher chemical richness emerge (limestones and marls in Monte Alegre-Ererê; basalts and diabases in Roraima, Pará, and Amapá States). In general, in the other areas, the current bioclimatic conditions, the characteristics of the source material, and the geoforms lead to the formation of deep and weathered soils [46].

## 2.2 MCT classification of tropical soils

The technical deficiencies in the highways of the Amazonia are the result—for the most part—of the use of local materials analyzed under the same experimental techniques based on research carried out in regions of low temperatures (temperate climate) and well-distributed rainfall throughout the year; this condition is totally different from the climate of the equatorial zone, which is characterized by intense climatic variations, high temperatures, and high rainfall incidence [47]. In addition, the traditional soil classification systems—TRB and USCS—disregard the essential evaluation of the mechanical and hydraulic attributes of geomaterials [48]. The use of those conventional methodologies for the classification of natural materials for application in road pavement results in the neglect of materials with potential properties for use in pavement layers when the object of study is tropical soils [49]. As an alternative, in the last three decades, several scientific studies have confirmed the importance of adopting the MCT system (nomenclature for Miniature, Compacted, Tropical)—created by Nogami and Villibor [50]—for the study of fine-grained tropical soils. The main purpose of the MCT methodology is to provide an understanding of the importance of rationalizing the use of tropical soils on highways, to reduce the costs of road work and its impacts on the environment, and establish the difference between lateritic soils (oxisols) and saprolitic (argisols) [51].

The MCT methodology has undergone several modifications over time to improve this classification system for practical road purposes, taking advantage of the lateritic tropical soils, which are abundant in many areas of Brazil [52–54]. It has a structured laboratory test program that is composed of the Mini-MCV and Mass Loss by Immersion with both tests carried out on compacted miniaturized samples. These tests yield the values of classification indexes c′ (determined from the deformability curve slope), d′ (angular coefficient of the dry side of the compaction curve, corresponding to 10 blows), e′ (laterization index), and Pi (mass loss by immersion, in %). These values are graphed and placed in a classification abacus (**Figure 11**), which performs a pedological separation of the materials.

On the other hand, it is worth noting that tropical soils also have horizons of occurrences consisting of, in addition to the fine-material content, a portion of coarse granulation formed by a gravel fraction of lateritic concretions (known as *piçarra*, in Brazilian Amazonia region). For this reason, Villibor and Alves [55, 56] suggested a new classification (termed G-MCT), considering the fine and coarse granulation contents (the distribution of the particle size is determined by the integral soil particle size tests and the G-MCT abacus), whereas the original MCT classification addresses only the percentage of fine-grained material. **Figure 12** shows the structure of the test program for the new G-MCT classification proposed by the authors (**Figure 12a**) and the classification of the coarse materials by particle size (**Figure 12b**). The connection of associated parameters in the G-MCT system aims to qualify the analysis of the results that determine the classification of the material, letting verify the feasibility of applying the material to the base and subbase layers.

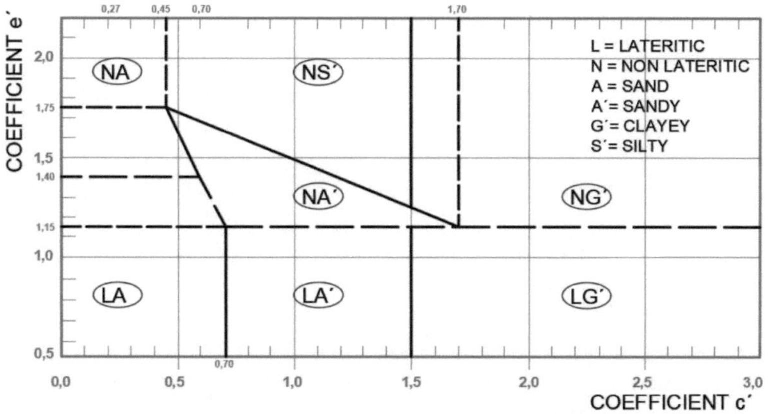


**Figure 11.**
*Graph for soil classification by the MCT method (adapted from [52]).*

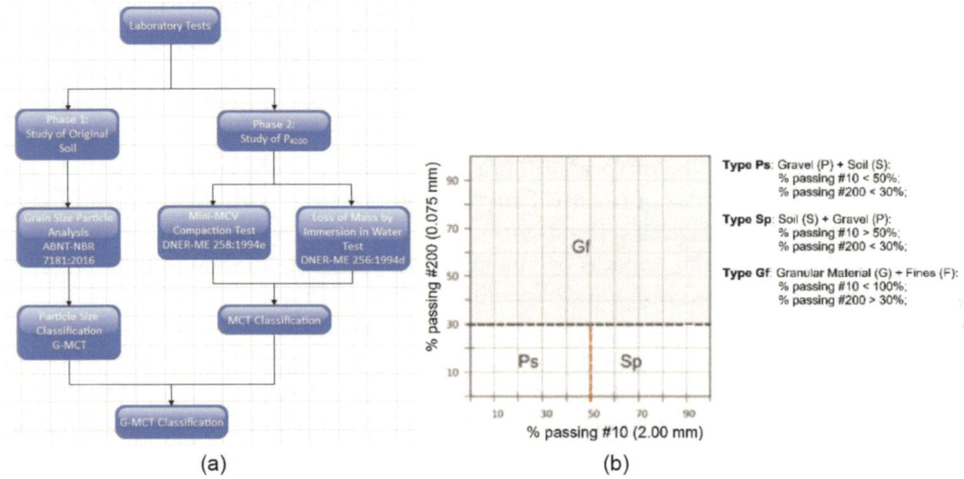


**Figure 12.**
*Test program for the G-MCT classification. (a) Flowchart; (b) classification (adapted from ref. [56]).*

Vertamatti [57] conducted the first study of the MCT methodology for Amazonian soils, evaluating the use of fine and coarse (lateritic concretions compatible with the gravel fraction size) texture of lateritic soils in airport projects in the Brazilian Amazonia. He found that the soils generally showed good stability against the water (rains) influence due to the fine plastics present, resulting in a cohesive structure responsible for the great durability of the base layer of airports runways without asphalt course, even under successive periods of heavy rains in the region.

Sant'Ana [58] studied 20 samples of lateritic soil in the State of Maranhão to compare the mini-MCV test and the "rapid disk" method, a test proposed by Nogami and Villibor [59] in which a fraction of soil passed through a #40 (0.42 mm) sieve is molded in a stainless steel ring, measured its contraction (after drying in an oven), and penetrated by a standard needle (after saturation in water), within the MCT methodology. The author found a better classification relationship of the lateritic soils with the results of the mini-MCV test.

Santos and Guimarães [60] evaluated the mechanical behavior of coarse lateritic concretion soils used in road paving in the city of Porto Velho, State of Rondônia, Brazil. These authors found high values of resilient modulus (between 350 and 600 MPa) and low values of permanent deformation of these soil types. In the study made by Barbosa [61], the MCT methodology was applied to soil samples collected in a deposit located in the city of Rio Branco, State of Acre, Brazil, for the production of a synthetic coarse aggregate of calcined clay (SCACC), and mixtures for base courses. Baia [47] and Baia et al. [51] performed a comparative analysis between the tests of the USCS and TRB and the MCT soil classification systems, with geomaterial samples collected from the rural zone (lateritic soil) and the margin of *várzeas*, in the city of Manaus, State of Amazonas, and found good results for the use of lateritic soil in road works.

Almeida [72] evaluated a tropical clayey soil collected from a deposit in the metropolitan region of the city of Manaus using the MCT methodology for the application of the material with a chemical additive (synthetic zeolite cement) as a solution for low traffic volume rural road.

Delgado [62] studied the application of an essentially clayey soil with a high plasticity index for use as a subballast layer in an expansive stretch of the Carajás Railway in the

Western region of the State of Maranhão. This soil would be discarded for the proposed purpose, considering the conventional standards of subballast selection, imported mostly from temperate climate countries. However, due to its tropical soil nature, the results obtained indicated high resilience modulus and low total permanent deformation values, showing that it is a material that, despite not meeting the criteria of traditional soil classification systems, would be adequate for use in the real field situations.

Although normalized by highway agencies in Brazil, there is still a lack of details about the MCT methodology of Brazilian tropical soils. It is important to continue efforts that yield new field and research data to define the geotechnical classification of these types of soils [56].

### 2.3 Stabilization of Amazonian soils with chemical additives

Before, during, and after floods caused by river floods or intense rains, the quality of the road infrastructure is essential [63]. During these events, pavement layers consisting of nonconsolidated (flexible) materials are more susceptible to erosion, while consolidated (rigid) materials are prone to failure when the lower layers are subject to erosion. Thus, evaluating pavements consisting of the *in situ* transformation of natural subgrade in a rigid base layer with cement and specific additives may be an option for regions with a lack of stone materials and flood risk. Some studies have shown the benefits, even in subgrades of low bearing capacity [64] of a reduction in the final thickness of the asphalt course.

The properties of cement-stabilized materials are strongly determined by the nature of the raw material used, which may be clay, silt, sand, or gravel. The type of soil influences the choice of stabilizer and controls the structural properties of the stabilized product. To a large extent, the variability of soil properties comes from the particle size distribution, arrangement of the particles, shape of the grains, and mineralogical composition [65].

Soil-cement structures are prone to hydraulic retraction, especially during the moisture loss caused by cement hydration or temperature changes. The accumulation of cracks caused by shrinkage can accelerate the damage to the pavement, the erosion processes, and the reduction of the strength and durability of the base layer. Conversely, the addition of synthetic zeolite (ZS) additive, together with cement, for *in situ* soil transformation modifies the cement hydration process on a nanometric scale, improving the formation processes of the crystalline microstructure, exchange ionization, adsorption, and immobilization of potentially harmful compounds in soils that, in a traditional approach, would need to be removed or discarded, at significant cost, making them relevant and suitable materials to be used in road construction [66] (**Figure 13**). ZS additives can increase the strength and stiffness of the soil–cement composites and improve the overall performance of the stabilized layers of pavement [67–69]. In the State of Amazonas, laboratory tests performed in clayey soils have revealed a gain in simple compressive strength (SCS) tests when adding ZS (RoadCem®) to its composition [70–72]. For field application, this additive is used at a low dosage (1.2–2.4 kg/m$^3$) [73]. The dosage of the additive can be increased based on the local conditions, such as the soil characteristics, the time of the opening of traffic, and the climatic conditions present during construction.

**Figure 14a** shows scanning electron microscopy (SEM) images of soil stabilized with 8.2% cement, without additive, and with 0.174% ZS additive (RoadCem®) (**Figure 14b**) for the submerged curing condition. The samples were cured for 28 days, and the products of cement hydration and pozzolanic reactions (cation

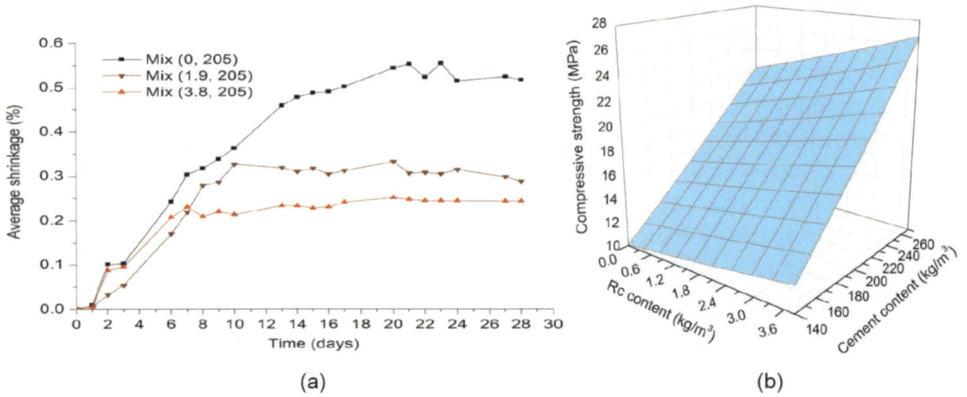


**Figure 13.**
*(a) Effect of synthetic zeolite on shrinkage reduction and (b) on SCS gain as a function of dosage (adapted from [73]).*

**Figure 14.**
*Scanning electron microscopy (SEM) images of stabilized soil with 8.2% cement: (a) without; and (b) with RoadCem® additive, for submerged curing at 28 days [72].*

exchange and flocculation) already took place, joining the flocculated clay particles. A denser and more complex structure was observed in the samples with additives, indicating a greater amount of cement hydration products. Ettringite crystals (calcium sulfate and hydrated aluminate, with a size of 1 μm) were formed in needle shape in samples with the ZS additive under both curing conditions.

## 2.4 Synthetic coarse aggregate of calcined clay (SCACC) in road paving

The execution of the base, subbase, and asphalt courses consumes a large volume of coarse aggregate. As explained in Item 1, there is a shortage of natural stone material (crushed stone or pebble) in the ASB, so there is a high cost of road paving construction services in this area.

The process of producing light expanded clay aggregates in a rotary kiln began in 1908 [74]. Depending on the clay mineral constituent of the raw material there are some types of clay that, when burned at temperatures below 800°C, do not generate light aggregates but only a calcined, coarse synthetic aggregate that is not expanded

[75]. This latter material is designated here as SCACC [76] or "burnt clay" and can be used both in pavement subbases and bases, and in asphalt coatings and surface treatments [77].

The Texas Highway Department established a classification system for synthetic clay aggregates [78]. Moore et al. [79] stated that the firing temperature to produce the calcined aggregate should in general lower than that of the expanded clay aggregate, only hot enough to completely dehydrate the clay, approximately 550–750°C, for approximately 15 min.

There are reports of the use of SCACC in Nigeria [80], English Guiana, Sudan, Australia [81], and Thailand [82]. Their results showed greater durability of the road pavings using SCACC, in addition to better skid resistance, than with the conventional aggregate.

In Brazil, research on light aggregates has been done since 1966 [83]. Fabrício [84] developed a mobile prototype plant for the manufacture of expanded clay aggregates or SCACC in road paving works in Amazonia. Campelo et al. [85] demonstrated—from laboratory tests—the technical, economic, and environmental feasibility of using SCACC in asphalt mixtures burning *in natura* clay from the potter pole of the cities of Iranduba and Manacapuru, State of Amazonas. Cabral et al. [86] reported the use of SCACC in structural concrete of Portland cement in the State of Amazonas, concluding that the mechanical strengths were similar to those with natural aggregates (pebble) with the same cement consumption.

Campelo et al. [23] studied SCACC because it is an alternative material that offers a competitive price in relation to the conventional aggregate (crushed stone or pebble), in addition to the fact that the raw material is a *várzea* clay, which is an abundant

| Material | Brazilian standard | Title | Acceptance Parameters |
|---|---|---|---|
| SCAAC | DNER—ME 225/94 | Synthetic aggregate of calcined clay—Pressure slaking test | Less than 6% |
| SCAAC | DNER@—ME 222/94 | Synthetic aggregate of calcined clay—Los Angeles abrasion test | Less than 35% |
| SCAAC, Pebble | NBR NM 53/2009 | Coarse aggregate—Determination of the bulk specific gravity, apparent specific gravity, and water absorption | Greater than 0.88 and 2.00 g/cm$^3$; less than 18%, respectively |
| Pebble | NBR NM 51/2001 | Coarse aggregate—Test method for resistance to degradation by Los Angeles machine | Less than 50% |
| SCAAC, Pebble, Sand, Filler | NBR NM 248/2003 | Aggregates—Sieve analysis of fine and coarse aggregates | Within granulometric range |
| SCAAC, Pebble | NBR 12583/1992 | Coarse aggregate—Coating to bituminous binder | Qualitative test (visual analysis) |
| Sand | NBR NM 52/2009 | Fine aggregate—Determination of the bulk specific gravity and apparent specific gravity | Greater than 1.60 and 2.60 g/cm$^3$, respectively |
| Filler | NBR NM 23/2001 | Portland cement and other powdered material—Determination of density | Greater than 3.00 g/cm$^3$ |

**Table 2.**
*Natural and synthetic aggregate characterization tests [23].*

mineral resource in the Amazon Basin. This aggregate can be used in the subbase, base, and asphalt courses of the pavement, in addition to the surface and deep drainage layers of the pavement. **Table 2** shows the technological properties that the SCACC must have to be used in pavement layers. Campos et al. stated that the SCACC can be calcined at the temperatures at which bricks and ceramic tiles are normally produced, that is, between 850 and 950°C. For the base or subbase courses, these temperatures may be lower. The asphalt mixtures made with SCACC offer a better fit between their components than those produced with rolled pebbles, favoring greater mechanical strength. The SCACC aggregate, when applied in asphalt mixtures, represents a more resistant structural skeleton than the pebble, although the latter is the most used in the State of Amazonas.

## 2.5 Stabilization of Amazonian soils with soil-emulsion mixtures

The execution of stabilization with asphalt emulsion consists of two stages, spreading and compaction [87]. According to a study by Klinsky [88], soil-emulsion stabilization can occur in the following combinations:

- Sand-Asphalt: Generates cohesion effect in materials with a fraction passing through the #200 sieve (0.074 mm) of between 5 and 12% and a plasticity index (PI) < 10%;
- Soil-Asphalt: Reduces capillarity and infiltrability in clay-silty and clay-sandy soils;
- Gravel-Bitumen: Provides cohesive effect in materials with a fraction passing through the #200 sieve of <12% and PI <10%.

In the study by Rebelo [89] with a soil sample from the city of Coari, State of Amazonas, it was demonstrated that after 7 days of curing, the addition of asphalt emulsion in the geomaterial provided increased resistance to the mixture. Sant'Ana [58], when evaluating soil samples from the Northwest region of the State of Maranhão, suggested specific guidelines and conditions for the acceptance of materials and dosages for asphalt stabilization. This author recommended determining the "optimum" content through its correlation with the SCS tests or tensile strength by diametral compression (TSDC) tests under the conditions of dry curing (air) and immersion for 7 days. The author considered immersed curing because he observed that specimens with 7 days of air curing showed more resistance, even without emulsion, though the same did not occur when applying immersion in steps that preceded the SCS test, as shown in **Figure 15**.

Soils to be stabilized with high levels of emulsifier should be discarded because they make the stabilization services unfeasible economically [58]. In addition, according to Ingles and Metclaf [90], excess binder impairs the interaction between the grains caused by the lubrication of the particles, thus decreasing the resistance of the mixture.

Baia [47], when evaluating the microstructure of soil samples extracted in the rural area of the city of Manaus, with and without the addition of asphalt emulsion (optimum content of 4%), found a certain volume increase (**Figure 16b**) of the solid-phase soil caused by the inclusion of the emulsion in the intergranular spaces. Similar behavior had been observed in soils of the city of Rio de Janeiro by Miceli Jr. [91], who concluded that the higher the binder content in the soil, the greater the volumetric expansion.

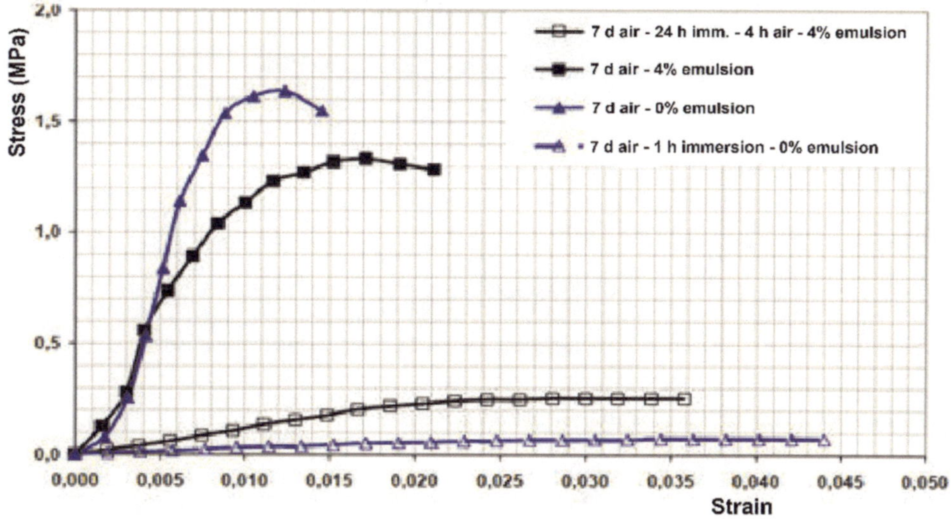


**Figure 15.**
*SCS test results of lateritic soil without emulsion, and with 4% emulsion, with 7 days of curing, immersed and non-immersed [58].*

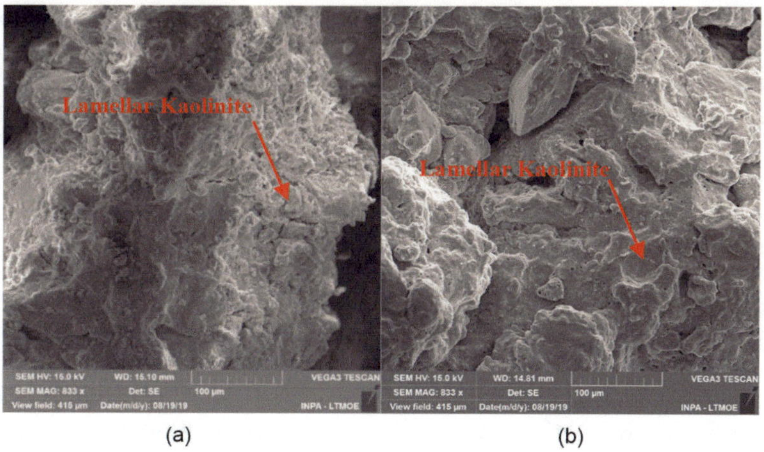


**Figure 16.**
*Scanning electron microscopy (SEM) images—magnified 500× (100 μm). (a) Natural soil; (b) soil with 4% RL-1C asphalt emulsion [47].*

## 2.6 Stabilization of Amazonian soils with ceramic waste

It is common to have a loss in the production process of red ceramics due to several factors, including the failure in the process of mixing and homogenization of the raw material *in natura*, the conformation, drying, and inadequate burning and transport of the product [92].

Waste ceramics are stacked in inappropriate places (**Figure 17**) within the pottery industry yard limits and may be a refuge for venomous and disease-transmitting animals [85]. This material is reused in the pottery industry itself [93], as a reducer of the plasticity of the *in natura* clay [94], or in base and subbase courses of road pavings [95]. According to Campelo et al. [94], the loss of burnt red ceramic products

**Figure 17.**
*Ceramic waste stacked in the pottery yards (photos: NS Campelo).*

(ceramic bricks and tiles) in the main production center of the State of Amazonas varied, on average, between 3 and 5% of production, though there were some ceramic industries in which this loss could reach 10%, which at that time would reach a total of 10,000 t/year.

Dias [95] evaluated the mechanical behavior of the typical soil (a yellow lateritic sandy silty-clay) of the city of Manaus, stabilizing it with ceramic waste to apply it in base and subbase courses. This author stated that the loss of red ceramic products (nonstructural and structural bricks, and tiles) was approximately 135 t/day of ceramic waste in the potter pole of the cities of Iranduba and Manacapuru. The author analyzed several soil-ceramic waste mixtures with different proportions, concluding that the mixture with the best performance to be used in the base or subbase course was 30% natural clayey soil, 30% sand, and 40% ceramic waste.

According to the study of Dias [95], based on the proportion of use of ceramic waste determined in the laboratory, it is then possible to construct a base or subbase course 65 km long, with 7 m of platform width, and 10 cm of thickness compacted, considering the bulk density of the ceramic waste as 1.81 $t/m^3$. Considering the small total length of the urban and rural roads of the interior cities typical of the Brazilian Amazonia (cities with fewer than 30,000 inhabitants in the urban area), this length of 65 km represents a considerable portion of the total of the existing roads in these cities; therefore, the reuse of this waste would bring environmental, economic, and human health gains.

## 2.7 Reinforced piled embankment in highways of the Brazilian Amazonia

Historically, wooden piles have been used in the Brazilian Amazon, especially since the peak of natural rubber exploitation (1870–1920), in which European companies—mostly of British origin—were responsible for the construction of the infrastructure (concessionaires of water, sewage, electricity, ports, urban transport by electric trams, in addition to the paving of urban and rural roads, buildings, bridges, etc.) in the cities of Manaus and Belém, the two largest capitals of the Brazilian Amazonia.

Until the end of the 1980s, it was possible to use wooden piles for not very high loads in the Brazilian Amazonia, but then, due to a series of restrictions imposed by environmental laws, it is now virtually impossible to use this type of material unless it comes from certified areas. However, it is possible to use other types of piles, such as precast concrete and on-site piles.

A reinforced piled embankment consists of a geogrid-reinforced landfill on a pile foundation; generally, one or more horizontal geosynthetic reinforcement layers are installed at the base of the landfill. This geotechnical solution can be used for the construction of a road or railroad, when a traditional construction method would require a long construction time or when the excavation of soil with low bearing capacity could affect buildings in the neighborhood or even result in a substantial residual settlement, making frequent maintenance of these works necessary [96]. This is an excellent option for sites with natural subgrade formed by thick layers of soils with low bearing capacity, in which it is uneconomical to purge this material, either due to the volume of excavation or the time to be spent in the service.

Unfortunately, although this type of solution had already been used in some road works in the Brazilian Amazonia, this was not documented, except in those found by Silva [97] (**Figure 18**) and Maccaferri [98]. Silva [97] showed the use of wooden piles in a road embankment on a natural subgrade consisting of a very soft clay layer, 35 m thickness, driving the piles with dimensions of 25 cm in diameter and 10 m in length in equal spacings of 1.60 m in the plan. This road embankment is adjacent to a trapezoidal earthen open channel, using a geotextile reinforcement layer above the pile cap of the piles.

## 2.8 Use of lateritic concretions in road Pavings in the Brazilian Amazonia

According to Swanson [99], the term *laterite* comes from the Latin word *later*, which means brick or tile, and was first suggested by Buchanan because the predominant color of laterite is red and it is often hard. Laterites arose from the decomposition of aluminous minerals by changes that seem peculiar to the tropics due to the action of the tropical forest on the soil. The laterite deposits are geographically restricted because they require tropical heat, intense rainfall, and luxuriant vegetation for their formation. They also require rainy and dry seasons and elevated plains on gently sloping terrain surfaces, which are not subject to appreciable erosion.

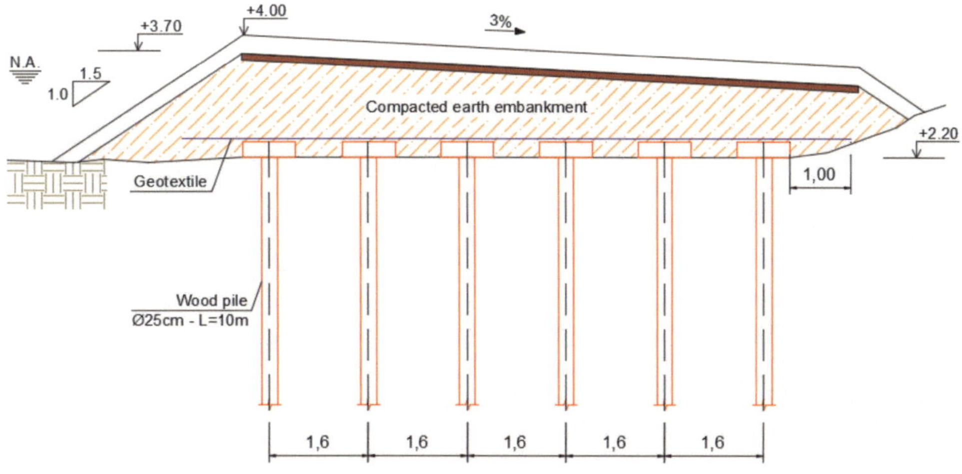


**Figure 18.**
*Use of wooden piles for the construction of a road embankment near the trapezoidal open channel (adapted from Ref. [97]).*

Much of the Amazonia landscape developed in lateritic terrain [100, 101]. The mature laterites are strongly leached from $SiO_2$ and alkalis but enriched in $Al_2O_3$ and $Fe_2O_3$ [102–104]. The laterites of the Amazon are masked by thick sandy-clayey to sandy-clayey latosols and/or by sedimentary cover [102, 105, 106]. Laterite profiles typical of the Amazon are illustrated in **Figure 19**, showing their main horizons [107, 105]. Tropical residual soils, especially lateritic deposits, which have been the subject of research in the Amazonia for use in road paving, can also be identified [47, 60, 108, 109] due to their peculiarities in comparison to sedimentary soils, as previously described. Keller et al. [110] stated that the materials commonly used in the Amazonia region for road paving include local laterite deposits but that despite being a hard and cemented material, they may still contain a high clay content. As a paving material, it is widely used for the structural layers of highways, but the first challenge is to locate the deposits with an adequate quantity and quality of the material. **Figure 20** shows some unusual uses of wood for the drainage of small watercourses (*igarapés*) in rural roads with low traffic volume in the Amazonia, such as rustic wooden bridges (locally called *pinguelas*) and "tubes" of drainage formed by hollow tree trunks. Washed coarse laterite (to remove fine grains) has also been used in

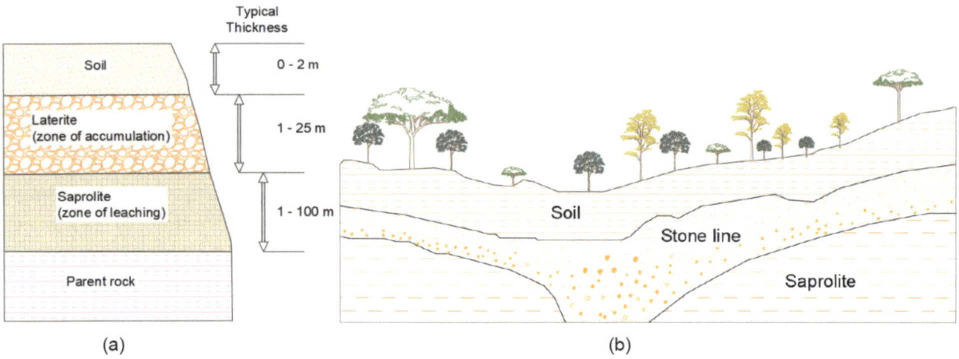


**Figure 19.**
*Typical profile of laterite from the Brazilian Amazonia and its main horizons: (a) adapted from Ref. [107]; (b) adapted from Ref. [105].*

**Figure 20.**
*Use of wood in the drainage of* igarapés *crossing roads with low traffic volume in the Brazilian Amazonia. (a) Wooden bridge (*pinguela*); (b) hollow wood "tube" [110].*

| Brazilian standard | Title | Acceptance parameters |
|---|---|---|
| DNER-ME 030/94 | Soils-determination of the silica-alumina and silica-sesquioxide relations of soils | $\leq 2$ |
| DNER-ME 172/2016 | Soils—determination of the California Bearing Ratio (CBR) using deformed and unhandled soil samples | Expansion $\leq 2\%$<br>CBR $\geq$ 80% (N > $5 \cdot 10^6$)<br>CBR $\geq$ 60% (N $\leq 5 \cdot 10^6$) |
| NBR-NM 248/2003 | Aggregates—Sieve analysis of fine and coarse aggregates | $P_{\#200} \leq 30\%$<br>$P_{\#200} \leq (2/3) \cdot P_{\#40}$ |
| DNER-ME 082/94 | Soils—determination of the plastic limit | $P_{\#}40$:<br>LL $\leq$ 40%; IP $\leq$ 15% |
| NBR-NM 51/2001 | Coarse aggregate—test method for resistance to degradation by Los Angeles machine | $\leq 65\%$ |

*Notes: $P_{\#200}$, $P_{\#40}$: percentage of particles passing through the #200 and #40 sieves, respectively; LL: liquid limit; PI: plasticity index.*

**Table 3.**
*Technical specifications for granulometrically stabilized base services using lateritic soil [111].*

asphalt mixtures. **Table 3** shows the requirements for the use of laterite as a natural coarse aggregate in base or subbase layers, according to specifications [111].

## 2.9 Concepts of unsaturated soil mechanics applied to tropical soils

Much of the tropical topsoil is subject to unsaturated conditions, that is, not all its voids are filled by the aqueous phase. This is true both in the *terra firme* (non-flooded regions) and in the *várzea* (lowlands) areas of the Amazonia when these are not subject to the flood pulse. This unsaturated condition means a different behavior of the soil massif in terms of shear strength, in comparison to that under the saturated soil condition [112].

Many engineering problems involve unsaturated soils. The construction of earth dams, highways, and airport runways uses compacted soils that are not saturated. An element of unsaturated soil can, therefore, be viewed as a mixture with two phases that reach equilibrium under applied stress gradients (i.e., soil particles and contractile skin) and two phases that flow under applied stress gradients (i.e., air and water) [113].

In unsaturated soils, the pore-water pressures are negative in relation to atmospheric conditions; this negative pore pressure is called matric suction when referring to the air pressure [114, 115]. As the soil approaches saturation, the pore-water pressure approaches the pore-air pressure. Therefore, the matric suction tends to zero, and there is a smooth transition to the stress state of saturated soil [114].

Fredlund and Rahardjo [116] reported that in recent years there is a better understanding of the role of negative pore pressure (or matric suctions) in increasing the shear strength of the soil, and that it is appropriate to perform an analysis of slope stability including the matric suction contribution.

Regarding slope stability, Ching et al. [117] reported that soil suction profiles play a significant role in the long-term stability of many natural slopes and steep cutting slopes. However, during or after periods of intense and prolonged rainfall, slope failures often occur because rains cause infiltration in the ground and reduce soil resistance because of matric suction loss. This augments the safety factors by considering the actual matric suctions of the soil, and therefore, they contribute substantially to the increases in shear strength [118, 119] and slope stability in unsaturated conditions [117, 114, 120, 121].

In tropical and subtropical areas, slope failures induced by rain are closely related to soil properties, slope geometry, groundwater position, and certain environmental factors, vegetation, and weathering effects [122]. Thus, the slopes are stable, with a high safety factor during dry periods, and tend to fail only during rainy periods [119].

## 3. Conclusions

The construction of highways in the Brazilian Amazonia is problematic because it faces nature-related, technical, economic, and environmental issues, which are interrelated.

The natural questions come from a range of origins, the main ones being geological-geotechnical, pedological, relief-related, and climate-related, and obviously cannot be gotten around given their territorial scope. The technical issues concern the natural subgrade, especially in floodplains (*várzeas*), formed by fine alluvial soils (silts and clays), which are plastic, impermeable, highly compressible, or expansive, present in the vast Amazon Basin. The economic issues are related to the costs of transporting stone material and lateritic soils—some with the presence of lateritic concretions (*picarras*)—to the construction sites since they occur in limited portions of the Amazonia (*terras firmes*). Environmental issues fall into a vicious circle since the construction of more highways tends to reach areas of virgin forest, which may be subject to new deforestation processes and other environmental impacts, affecting the rich fauna and flora of the region, in addition to native communities. Luckily, Brazilian environmental laws have become increasingly rigid, requiring in-depth studies of environmental impacts and public hearings before the construction of roadworks is licensed.

## Acknowledgements

The authors would like to thank CAPES for the grant of a scholarship for an academic master's degree to the Civil Engineer Daniel Jardim Almeida.

## Author details

Nilton de Souza Campelo[1*], Arlene Maria Lamêgo da Silva Campos[1], Marcos Valério Mendonça Baia[1], Daniel Jardim Almeida[1], Raimundo Humberto Cavalcante Lima[2], Danielly Kelly dos Reis Dias[1], Júlio Augusto de Alencar Júnior[3] and Mário Jorge Gonçalves Santoro Filho[4]

1 Graduate Program in Civil Engineering (PPGEC), Federal University of Amazonas (UFAM), Manaus, AM, Brazil

2 Graduate Program in Geosciences (PPGGEO), Federal University of Amazonas (UFAM), Manaus, AM, Brazil

3 Department of Civil Engineering (DEC), Federal University of Pará (UFPA), Belém, PA, Brazil

4 STP Engineering Projects & Constructions, Manaus, AM, Brazil

*Address all correspondence to: ncampelo@ufam.edu.br

## References

[1] Brazilian Institute of Geography and Statistics – IBGE. Geosciences: Environmental Information. Rio de Janeiro, Brazil. 2021 [in Portuguese]

[2] Food and Agriculture Organization of the United Nations—FAO. The State of Forests in the Amazon Basin, Congo Basin and Southeast Asia A report prepared for the Summit of the Three Rainforest Basins Brazzaville. Republic of Congo: FAO; 2011

[3] Luz HR, Martins TF, Muñoz-Leal S, Costa FB, Gianizella SL, Faccini JLH, et al. Ticks from the Brazilian Amazon: Species, distribution and host-relations. In: Mikkola HJ, editor. Ecosystem and Biodiversity of Amazonia. Chapter 3. London: IntechOpen; 2021. pp. 33-66. DOI: 10.5772/intechopen.94862

[4] Motta RS. The Economics of Biodiversity in Brazil: The Case of Forest Conversion. Rio de Janeiro, Brazil: Institute for Applied Economic Research (IPEA); 1996. p. 25

[5] UNEP—United Nations Environment Programme. Megadiverse Brazil: Giving Biodiversity an Online Boost. Report. Brazil: UNEP; 2019

[6] Shepherd, GJ. Terrestrial plants. In: Lewinsohn TM, editor. Assessment of the State of Knowledge of Brazilian Biodiversity. Biodiversity Series. Elsiver Brazil 15(2): 145-192 [in Portuguese]

[7] Forzza RC et al. Catalog of Plants and Fungi from Brazil. Vol. 1. Rio de Janeiro: Rio de Janeiro Botanical Garden Research Institute; 2010 [in Portuguese]

[8] MMA—Ministry of the Environment. Chapter II: The Status of Brazilian Biological Diversity. State-of-the-Art of the Knowledge of Biological Diversity. Brazil, Brasilia: First National Report to the Convention on Biological Diversity; 1997

[9] Peel MC, Finlayson BL, McMahon TA. Updated world map of the Koppen-Geiger climate classification. Hydrology and Earth System Sciences. 2007;**11**:1633-1644

[10] Nobre CA, Obregón GO, Marengo JA, Fu R, Poveda G. Characteristics of Amazonian climate: Main features. In: Keller M, Bustamante M, Gash J, Silva DP, editors. Amazonia and Global Change. Geophysical Monograph Series. Vol. 186. Washington, D.C.: American Geophysical Union Books; 2009. pp. 149-162

[11] Dias MAFS. Forest and rainfall interactions in the Amazon Basin. Terrae. 2008;**1**:46-53

[12] Sorí R, Marengo JA, Nieto R, Drumond A, Luis Gimeno I. The atmospheric branch of the hydrological cycle over the Negro and Madeira River Basins in the Amazon Region. Water. 2018;**10**(738):1-29. DOI: 10.3390/w10060738

[13] Junk WJ, Piedade MTF, Schöngart J, Cohn-Haft M, Adeney JM, Wittmann F. A classification of major naturally-occurring Amazonian lowland wetlands. Wetlands. 2011;**31**:623-640. DOI: 10.1007/s13157-011-0190-7

[14] Junk W, Bayley PB, Sparks RE. The flood pulse concept in river-floodplain systems. In: Dodge DP, editor. Proceedings of the International Large River Symposium (LARS). Vol. 106. Canada: Canadian Special Publication of Fisheries and Aquatic Sciences; 1989. pp. 110-127

[15] Parolin P, Ferreira LV, Albernaz LKM, Almeida SS. Tree species distribution in *várzea* forests of Brazilian Amazonia. Folia Geobotanica. 2004;**39**: 371-383

[16] Souza ES, Campelo NS, Lima RHC, Aguiar RL. Geotechnical characterization and modelling of the "Fallen Lands" phenomenon in the Amazon environment. Global Journal of Engineering and Technology Advances. 2021;**09**(03):122-132. DOI: 10.30574/gjeta.2021.9.3.0168

[17] Cochrane TT, Sánchez PA. Land resources, soils and their management in the Amazon region: A state of knowledge report. In: Hecht SB, Nores GA, Sánchez PA, Spain JM, Toenniessen G, editors. International Conference on Amazonian Agriculture and Land Use Research; 1980. Cali: Colombia; 1982. pp. 137-209

[18] Rabelo RM, Pereira GCN, Valsecchi J, Magnusson WE. The role of river flooding as an environmental filter for Amazonian butterfly assemblages. Frontiers in Ecology and Evolution. 2021;**9**:693178. DOI: 10.3389/fevo.2021.693178

[19] Sioli H. General features of the limnology of Amazônia. Internationale Vereinigung für theoretische und angewandte Limnologie: Verhandlungen. 1964;**15**(2): 1053-1058. DOI: 10.1080/03680770.1962.11895647

[20] Conciani W, Campelo NS, BJI C, Souza PR, Barbiero N. Monitoring of suction pressure in a soil profile in the mid-Solimões region, in the construction of the Coari-Manaus gas pipeline. In: XIV Brazilian Congress of Soil Mechanics and Geotechnical Engineering (COBRAMSEG 2008); 23–26 August 2008; Búzios, Brazil. São Paulo: ABMS; 2008. pp. 1432-1439 [in Portugueses]

[21] Quesada CA, Lloyd J, Anderson LO, Fyllas NM, Schwarz M, Czimczik CI. Soils of Amazonia with particular reference to the RAINFOR sites. Biogeosciences. 2011;**8**:1415-1440. DOI: 10.5194/bg-8-1415-2011

[22] Oliveira NT. Highways in the Amazon: A geopolitical discussion. Confins-Franco-Brazilian Journal of Geography. 2019;**501**. DOI: 10.4000/confins.21176 [in Portuguese]

[23] Campelo NS, Campos AMLS, Aragão AF. Comparative analysis of asphalt concrete mixtures employing pebbles and synthetic coarse aggregate of calcined clay in the Amazon region. International Journal of Pavement Engineering. 2017:1-12. DOI: 10.1080/10298436.2017.1309199

[24] Campos A, Campelo N, Alencar J Jr. Utilization of synthetic coarse aggregate of calcined clay in asphalt mixtures in the Amazon Region. In: Proceeding of the International Conference on Soil Mechanics and Geotechnical Engineering (19th ICSMGE). Seoul, Korea: ICSMGE; 17-22 September 2017. pp. 321-324

[25] Vilela T, Harb AM, Bruner A, Arruda VLS, Ribeiro V, Alencar AAC, et al. A better Amazon road network for people and the environment. PNAS. 2020;**117**(13):7095-7102. DOI: 10.1073/pnas.1910853117

[26] Pfaff A, Barbieri A, Ludewigs T, Merry F, Perz S, Reis E. Road impacts in Brazilian Amazonia. Geophysical Monograph Series. 2009;**186**:101-116. DOI: 10.1029/2008GM000737

[27] Nepstad D, Carvalho G, Barros AC, Alencar A, Capobianco JP, Bishop J, et al.

Road paving, fire regime feedbacks, and the future of Amazon forests. Forest Ecology and Management. 2001;**154**: 395-407

[28] Barber CP, Cochrane MA, Souza CM Jr, Laurance WF. Roads, deforestation, and the mitigating effect of protected areas in the Amazon. Biological Conservation. 2014;**177**:203-209. DOI: 10.1016/j.biocon.2014.07.004

[29] Alves DS. Space-time dynamics of deforestation in Brazilian Amazônia. International Journal of Remote Sensing. 2002;**23**(14):2903-2908. DOI: 10.1080/01431160110096791

[30] CNT—National Transport Confederation. Transport Yearbook CNT—Consolidated Statistics. Brasília, Brazil: CNT; 2021

[31] Reis NJ, Almeida ME, Riker SL, Ferreira AL. Geology and Mineral Resources of the State of Amazonas: Explanatory Text of Geological Maps and Mineral Resources of the State of Amazonas at Scale 1:1.000.000. CPRM—Geological Survey of Brazil: Manaus; 2006

[32] Almeida FFM de, Martin FC, Ferreira EO, Furque G. Tectonic Map of South America. 1:5,000,000. Explanatory Note: Commission for the Geological Map of the World. DNPM/MME: Brasília, Brazil.; 1978. 21 p.

[33] Castro DL, Fuck RA, Phillips JD, Vidotti RM, Bezerra FHR, Dantas EL. Crustal structure beneath the Paleozoic Parnaíba Basin revealed by airborne gravity and magnetic data, Brazil. Tectonophysics. 2014;**614**:128-145. DOI: 10.1016/j.tecto.2013.12.009

[34] Park E, Latrubesse EM. Modeling suspended sediment distribution patterns of the Amazon River using MODIS data. Remote Sensing of Environment. 2014;**147**:232-242. DOI: 10.1016/j.rse.2014.03.013

[35] UNEP—United Nations Environment Programme, Barthem RB, Charvet-Ameida P, LFA M, Lanna AE. Amazon Basin, GIWA Regional Assessment 40b. Kalmar, Sweden: University of Kalmar; 2004

[36] McClain ME, Naiman RJ. Andean influences on the biogeochemistry and ecology of the Amazon River. Bioscience. 2008;**58**(4):325-338

[37] Santos RE, Pinto-Coelho R, Fonseca R, Simões NR, Zanchi FB. The decline of fisheries on the Madeira River, Brazil: The high cost of the hydroelectric dams in the Amazon Basin. Fisheries Management and Ecology. 2018:1-12. DOI: 10.1111/fme.12305

[38] Sioli H. The Amazon: Limnology and Landscape Ecology of a Mighty Tropical River and its Basin (Monographiae Biologicae Series). Dordrecht: Springer; 1984. 800 p. DOI: 10.1007/978-94-009-6542-3

[39] Gibbs RJ. The geochemistry of the Amazon River system. Part I. The factors that control the salinity and the composition and concentration of the suspended solids. Geological Society of America Bulletin. 1967;**78**:1203-1232

[40] Rebouças AC. Groundwater. In: Rebouças AC, Braga B, Tundisi JG (editors). Freshwater in Brazil: Ecological Capital, Use and Conservation. São Paulo: Editora Escrituras; 1999, cap. 4, p. 117-150 [in Portuguese]

[41] Silva MSR, Miranda SAF, Domingos RN, Silva SLR, Santana GP. Classification of Amazonian Rivers: A strategy for the preservation of these

resources. Holos Environment. 2013; **13**(2):163-174 [in Portuguese]

[42] Sombroek WG. Amazon soils: a reconnaissance of the soils of the Brazilian Amazon region [thesis]. Wageningen: Landbouwhogeschool Wageningen University; 1966

[43] Sombroek WG. Soils of the Amazon. Region. In: Sioli H, editor. The Amazon (Monographiae Biologicae). Dordrecht: Springer; 1984. pp. 521-535. DOI: 10.1007/978-94-009-6542-3

[44] Plácido CG Jr. Distribution and chemical characterization of soil fertility in the State of Amazonas [thesis]. Manaus: Federal University of Amazonas; 2007 [in Portuguse]

[45] Schaefer CEGR, Lima HN, Vale JF Jr, JWV M. Land use and landscape changes in the Amazon: scenarios and reflections. Bulletin of the Museu Paraense Emílio Goeldi (Earth Science Series). 2000;**12**: 63-104 [in Portuguese]

[46] Lima HN. Genesis, chemistry, mineralogy and micromorphology of soils in the Western Amazon [thesis]. Viçosa, MG: Federal University of Viçosa, Brazil; 2001 [in Portuguese]

[47] Baia MVM. Physical stabilization of tropical soils for low-cost pavement bases in the metropolitan region of Manaus [thesis]. Manaus: Federal University of Amazonas (UFAM); 2019 [in Portuguese]

[48] Nogami JS, Villibor DF. Use of lateritic fine-grained soils in road pavement base courses. Geotechnical and Geological Engineering. 1991;**9**(3–4):167-182

[49] Balbo JT. Asphalt Paving: Materials, Projects and Restoration. Oficina dos Textos: São Paulo; 2007. 558 p [in Portuguese]

[50] Nogami JS, Villibor DF. A new soil classification for road use. In: Brazilian Symposium on Tropical Soils in Engineering. Rio de Janeiro: COPPE/UFRJ; 1981 [in Portuguese]

[51] Baia MVM, Almeida DJ, Souza ES, Barbosa EP, Araújo FAS, Lima AOP, et al. Comparative analysis of methodologies for geotechnical classification of two soils from the margin of water bodies in the central zone of Manaus, AM. In: Brazilian Congress of Soil Mechanics and Geotechnical Engineering (XVIII COBRAMSEG); Salvador. São Paulo: ABMS; 2018 [in Portuguese]

[52] Nogami JS, Villibor DF. Low Cost Paving with Lateritic Soils. Vilibor: São Paulo; 1995. 240 p. [in Portuguese]

[53] Villibor DF, Nogami JS, Beligni M, Cincerre JR. Pavements with Lateritic Soils and Maintenance Management of Urban Roads. São Paulo: ABPv/Federal University of Uberlândia (UFU), Brazil; 2000. 138 p [in Portuguese]

[54] Villibor DF, Nogami JS. Economic Pavements: Technology for the Use of Lateritic Fine Soils. Arte & Ciência: São Paulo; 2009. 292 p [in Portuguese]

[55] Villibor DF, Alves DML. Classification of fine and coarse-grained tropical soils. Revista Pavimentação. 2017;**XII**(43):17-37 [in Portuguese]

[56] Villibor DF, Alves DML. Low Cost Pavement for Tropical Regions: Design and Construction: New Considerations. Tribo da Ilha: Florianópolis; 2019. 544 p [in Portuguese]

[57] Vertamatti E. Contribution to geotechnical knowledge of soils in the

Amazon based on investigation of airports and MCT and Resilient methodologies [thesis]. São José dos Campos, Brazil: Instituto Tecnológico da Aeronáutica (ITA); 1988 [in Portuguese]

[58] Sant'Ana WC. Contribution to the Study of soil emulsion in pavements of low-traffic highways for the state of Maranhão [thesis]. São Paulo: University of São Paulo (USP), Brazil; 2009 [in Portuguese]

[59] Nogami JS, Villibor DF. Expedited identification of MCT classification groups for tropical soils. In: Proceedings of the Brazilian Congress of Soil Mechanics and Foundation Engineering (X COBRAMSEF): Foz do Iguaçu, Brazil. Vol. 4. São Paulo: ABMS; 1994. pp. 1293-1300 [in Portuguese]

[60] Santos GPP, Guimarães ACR. Contribution to the knowledge of the mechanical behavior of lateritic soils used in pavements in Southwest Amazonia. In: Transport Teaching and Research Congress. Belo Horizonte: National Association for Research and Education in Transport—ANPET; 2011 [in Portuguese]

[61] Barbosa VHR. Study of soils in acre for the production of calcined aggregates and mixtures for paving bases [thesis]. Rio de Janeiro: Military Institute of Engineering (IME); 2017 [in Portuguese]

[62] Delgado BG. Analysis of the deformability of a tropical soil from the West of Maranhão as a Sub-ballast material in the Railroad Carajás [thesis]. Ouro Preto, Brazil: Federal University of Ouro Preto (UFOP); 2012 [in Portuguese]

[63] Gersonuis B, Egyed C. Macro-economic Effects of Using the PowerCem Technology on Road Infrastructure in Flood Risk Areas. Paris, France: UNESCO-IHE Report; 2012

[64] Kolias S, Kasselouri-Rigopoulou V, Karahalios A. Stabilisation of clayey soils with high calcium fly ash and cement. Cement and Concrete Composites. 2005; **27**(2):301-313. DOI: 10.1016/j.cemconcomp.2004.02.019

[65] Molenaar A. Cohesive and Non-cohesive Soils and Unbound Granular Materials for Bases and Sub-Bases in Roads. Lecture Notes: Delft University of Technology; 2010

[66] Nogueira LA, Crisóstomo PHS, MPS S, Frota CA. Evaluation of the geotechnical behavior of soils from forest areas in the Amazon mixed with Portland cement and Roadcem®. In: Annual Paving Meeting (44a RAPv); 18-21 August. Brazil: Foz Do Iguaçu; 2015 [in Portuguese]

[67] Wu P. Cement stabilized materials with use of roadcem additive [thesis]. Delft: Delft University of Technology; 2015

[68] Marjanovic P, Egyed CE, de La Roij P, de La Roij R. The Road to the Future: Manual for Working with RoadCem. Vol. 2. Moerdijk, Netherlands: PowerCem Technologies BV; 2009

[69] Mutepfa WAT. Laboratory evaluation of the effect of cement concentration, water salinity and the RoadCem additive on kalahari soil strength [thesis]. Botswana: University of Botswana; 2010

[70] Campelo N. Use of RoadCem® Additive for Stabilization of Low Bearing Capacity Subgrades, for Road Use. Brazil: Pavement Laboratory (LPAV)—Internal report (restricted use—not published); 2012 [in Portuguese]

[71] Nogueira LA, Figueiredo PVC, Barbosa CA, Frota CA. Evaluation of the Tensile Strength of Portland-Roadcem® Soil-cement Mixture, Brazilian Congress of Soil Mechanics and Geotechnical Engineering (COBRAMSEG 2016); 19–22 outubro 2016; Belo Horizonte, Brazil. São Paulo: ABMS; [in Portuguese]

[72] Almeida DJ. Stabilization of tropical clayey soil with synthetic zeolite cement as a paving solution in Amazonas [thesis]. Manaus: Federal University of Amazonas; 2018 [in Portuguese]

[73] Vu BT, Ngo TTH, Nguyen QD, Ngo AQ, Ho LS. Study on Cement-Treated Soil with RoadCem Additive in Construction of Rural Roads: A Case Study in Viet Nam. Singapore: Springer; 2019

[74] ESCSI—Expanded Shale Clay and Slate Institute. Lightweight Concrete: History, Applications, Economics. Washington, DC: ESCSI; 1971

[75] Riley CM. Relation of chemical properties to the bloating of clays. Journal of the American Ceramic Society. 1951;**34**(4):121-128

[76] Campelo NS. Potential for the use of Coarse Synthetic Calcined Clay Aggregate (AGSAC) in Civil Construction and Road Infrastructure Works in the Amazon Region. Uberlândia, Brazil: Lecture given at the Federal University of Uberlândia (UFU); 2006

[77] Moore WM. Fired-Clay Aggregates for Use in Flexible Bases. Research Report Number 81–12. College Station: Texas A&M University; 1969

[78] Ledbetter WB, Gallaway BM, Moore WM, Buth E. A Recommended Synthetic Coarse Aggregate Classification System. Special Report. College Station: Texas A&M University; 1969

[79] Moore WM, Van Pelt RS, Scrivner FH, Kunze GW. Suitability of Synthetic Aggregates Made from Clay-Type Soils for Use in Flexible Base. Research Report Number 81–5. College Station: Texas A&M University; 1968

[80] Grainger GD. The Production of a Low-Grade Aggregate from Black Cotton Soil by Heat Treatment. London: Road Research Laboratory; 1951. pp. 1-5

[81] Whyte BW. Method of Burning Soil for Road Construction Employed by the Public Works Department, British Guiana. London: Road Research Laboratory; 1951. pp. 6-15

[82] Bunnag S, Lerdhirunwong B. Application of calcined clay as aggregate for asphalt pavement and surface treatment. In: Australian Road Research Board Conference (ARRB 13), v. 13, part 3, Construction, Pavements and Materials. Adelaide: ARRB; 1986. pp. 214-219

[83] DNER—National Department of Highways of Brazil. Feasibility Research of the Implantation of an Expanded Clay Factory in the Amazon Region—Final Report. Rio de Janeiro: IPR; 1981 [in Portuguese]

[84] Fabrício JM. Development of a clay aggregate mobile plant. In: Brazilian Annual Pavement Congress (21st RAPv). Vol. 1. Salvador: Brazilian Pavement Association (ABPv); 1986. pp. 150-188 [in Portuguese]

[85] Campelo NS, Nogueira LMG, Lopes ACR, Silva JFP. Production of Synthetic Coarse Aggregate of Calcined Clay in Pottery Pole of Iranduba City for Employing in Road Infrastructure and Civil Construction in State of Amazonas.

Study of Technical, Economic, Environmental and Commercial Viability—EVTEC. State of Amazonas Govern Program for Support to Research in Companies. PAPPE/FAPEAM/FINEP: Manaus; 2005 [in Portuguese]

[86] Cabral EM, Vasconcelos RP, Vieira RK, Campelo NS, Vieira AK, Silva CC, et al. Calcined clay applied in concrete. In: Thomas S, Sebastian M, George A, Weimin Y, editors. Advances in Materials Science, Recycling and Reuse of Materials and their Products. Vol. 3. London: CRC Press; 2013. pp. 35-52

[87] ABEDA—Brazilian Association of Asphalt Distributors. Basic Manual of Asphalt Emulsions. 2nd ed. Rio de Janeiro: ABEDA; 2010. 144 p [in Portuguese]

[88] Klinsky LM. Pavement Base Courses. Rio de Janeiro: Brazilian Pavement Association (ABPv); 2021. 475 p. [in Portuguese]

[89] Rebelo EP. Study of soil-emulsion mixture for the Urucu Region (Coari-AM) [thesis]. Manaus: Federal University of Amazonas (UFAM); 2009 [in Portuguese]

[90] Ingles OG, Metcalf JB. Soil Stabilization, Principles and Practice Description. Sydney: Butterworths; 1972. 374 p

[91] Miceli G Jr. Behavior of soils in the state of rio de janeiro stabilized with asphalt emulsion [thesis]. Rio de Janeiro: Military Institute of Engineering (IME); 2006 [in Portuguese]

[92] EC—European Commission. Reference Document on Best Available Techniques in the Ceramic Manufacturing Industry. Brussels: EC—European Commission; 2007. 260 p

[93] Silva HT, Guimarães LS, Dutra FA, Martins DC, Tolentino DS Jr, ASV C, et al. Reuse of red ceramic waste in the production of concrete for civil construction. Research, Society and Development. 2021;**10**(12):1-10. DOI: 10.33448/rsd-v10i12.20967

[94] Campelo NS, Morais MR, Aragão AF, Cabral EM, Rebelo EP, Pinheiro SC, et al. Study of the use of burnt ceramic waste ("chamote") from the pottery pole of the municipalities of Iranduba and Manacapuru—Amazon State, as an additive in the manufacture of tiles. Cerâmica Industrial. 2005;**10**:1-3 [in Portuguese]

[95] Dias DKR. Reuse of ceramic waste from the pottery center of iranduba and manacapuru municipalities for use as a constituent element of the base and sub-base of pavement [thesis]. Manaus: Federal University of Amazonas (UFAM); 2016 [in Portuguese]

[96] Van Eekelen SJM, Brugman MHA. Basal Reinforced Piled Embankments—The Design Guideline. London: CRC Press; 2016. 156 p. DOI: 10.1201/9781315389806

[97] Silva KRM. The implementation of civil works and sanitation in the Una basin, in Belém, State of Pará, and the constraints related to the geological and geotechnical characteristics [thesis]. Belém: Federal University of Pará (UFPA); 2004 [in Portuguese]

[98] Maccaferri. Case History: Ponte Igarapé-Miri. Pará State, Brazil: Maccaferri; 2016; 2 p [in Portuguese]

[99] Swanson CO. The origin, distribution and composition of laterite. Journal of the American Ceramic Society. 1923;**6**:1248-1260. DOI: 10.1111/j.1151-2916.1923.tb17709.x

[100] Espíndola CR, Daniel LA. Laterite and lateritic soils in Brazil. FATEC-SP Technical Bulletin—BT/24. 2008;**5**:21-24 [in Portuguese]

[101] Costa ML. The two most important lateritization cycles in the Amazon region and their paleoecological importance. Proceedings of the Brazilian Academy of Sciences. 2001;**73**(3): 461-462 [in Portuguese]

[102] McNeil M. Lateritic soils in distinct tropical environments: Southern Sudan and Brazil. In: Farvar MT, Milton JP, editors. The Careless Technology: Ecology and International Development. Garden City: The Natural History Press; 1972. pp. 591-608

[103] Herbillon AJ, Nahon D. Laterites and laterization processes—Chapter 22. In: Stucki JW, Goodman BA, Schwertmann U, editors. Iron in Soils and Clay Minerals. NATO ASI Series (Series C: Mathematical and Physical Sciences). Vol. 217. Dordrecht: Springer; 1988. pp. 779-796. DOI: 10.1007/978-94-009-4007-9_22

[104] Prasad TK, Parthasarathy GR. Laterite and laterization—A geomorphological review. International Journal of Science and Research (IJSR). 2018;**7**(4):578-583. DOI: 10.21275/ART20181444

[105] Costa M. Geological aspects of Amazonian laterites. Revista Brasileira de Geociencias. 1991;**21**(2):146-160

[106] Tardy Y. Pétrologie dês Laterites et dês sols Tropicaux. Paris: Masson; 1993. 459 p

[107] Dagenais PJ, Poling GW. Acid rock drainage in the tropics: an analysis of the acid-generating characteristics of laterite/saprolite deposits and a comparison to hardrock deposits. In: Goldsack D, Belzile N, Yearwood P, Hall G, editors. Proceedings of the Sudbury '99, Mining and the Environment II, V. 1, Sudbury, Canada, 13-15 September. Sudbury, Canada: Laurentian University; 1999. pp. 79-88

[108] Almeida DJ, Oliveira FHL, Baia MVM, Campelo NS. Development of semi-rigid pavement adapted for the Amazon region with cement, synthetic zeolite and subgrade soil. Revista Militar de Ciência e Tecnologia. 2021;**38**(3): 14-26 [in Portuguese]

[109] Campelo NS, Carvalho JS, Carneiro BJI, Albiero JH. Verification of stability of slopes subject to gullies, in the city of Manaus. In: Brazilian Congress of Engineering and Environmental Geology (X CBGEA); 25-28 agosto 2002; Ouro Preto, Brazil. Vol. 1. São Paulo: ABMS; 2002 [in Portuguese]

[110] Keller G, Sherar J, Zweede J. An Amazon Basin forest roads manual-overview. In: Transportation Research Board (TRB). 11th International Conference on Low-Volume Roads. Pittsburgh, PA: TRB; 2015

[111] DNIT 098/2007—ES. Pavement—Granulometrically Stabilized Base Using Lateritic Soil—Service Specification. Rio de Janeiro, IPR; 2007 [in Portuguese]

[112] Fredlund DG, Morgenstern NR, Widger RA. The shear strength of unsaturated soils. Canadian Geotechnical Journal. 1978;**15**(3):313-321

[113] Fredlund DG, Morgenstern NR. Stress state variables for unsaturated soils. Journal of the Geotechnical Engineering Division (ASCE). 1977;**103** (GT5):447-466

[114] Fredlund DG, Rahardjo H. Theoretical context for understanding

unsaturated residual soil behavior. In: Proceedings, 1st International Conference on Geomechanics in Tropical Lateritic and Saprolitic Soils. Vol. 1. Brasília, Brazil; Sao Paulo: ABMS; 1985. pp. 295-305

[115] Rahardjo H, Lim TT, Chang MF, Fredlund DG. Shear-strength characteristics of a residual soil. Canadian Geotechnical Journal. 1995;**32**: 60-77

[116] Fredlund DG, Rahardjo H. Soil Mechanics for Unsaturated Soils. New York: John Wiley & Sons Inc.. 544 p

[117] Ching RKH, Sweeney DJ, Fredlund DG. Increase in factor of safety due to soil suction for two Hong Kong slopes. In: Proceedings of the Fourth International Symposium on Landslides. Toronto, Canada: Canadian Geotechnical Society; 1984. pp. 617-623

[118] Taha MR, Hossain MK, Mofiz SA. Effect of suction on the strength of unsaturated soil. In: Advances in Unsaturated Geotechnics (Geo-Denver 2000). Denver, Colorado, United States: American Society of Civil Engineers; 2000. DOI: 10.1061/40510(287)14

[119] Rahardjo H, Kim Y, Satyanaga A. Role of unsaturated soil mechanics in geotechnical engineering. International Journal of Geo-Engineering. 2019;**10**:8. DOI: 10.1186/s40703-019-0104-8

[120] Babu GLS, Murthy DSN. Reliability analysis of unsaturated soil slopes. Journal of Geotechnical and Geoenvironmental Engineering. 2005; **131**:1423-1428. DOI: 10.1061/(ASCE) 1090-0241(2005)131:11(1423)

[121] Zhang LL, Fredlund DG, Fredlund MD, Wilson GW. Modeling the unsaturated soil zone in slope stability analysis. Canadian Geotechnical Journal. 2014;**51**:1384-1398. DOI: 10.1139/cgj-2013-0394

[122] Rahardjo H, Nio AS, Leong EC, Song NY. Effects of groundwater table position and soil properties on stability of slope during rainfall. Journal of Geotechnical and Geoenvironmental Engineering. 2010;**136**:1555-1564. DOI: 10.1061/(ASCE)GT.1943-5606.0000385

Chapter 6

# Challenges in the Construction of Highways in the Brazilian Amazonia Environment: Part II – Engineering Solutions

*Nilton de Souza Campelo, Arlene Maria Lamêgo da Silva Campos, Marcos Valério Mendonça Baia, Daniel Jardim Almeida, Raimundo Humberto Cavalcante Lima, Danielly Kelly dos Reis Dias, Júlio Augusto de Alencar Júnior and Mário Jorge Gonçalves Santoro Filho*

## Abstract

The Chapter "Challenges in the construction of highways in the Brazilian Amazonia environment: Part I: Identification of Engineering Problems" dealt with the identification of Engineering problems concerning the implementation of road infrastructure in the Brazilian Amazon. This present chapter deals with Engineering solutions to overcome the obstacles mentioned in that previously chapter, focusing on alternatives that contemplate the use of the synthetic coarse aggregate of calcined clay (SCACC), recycled materials from ceramic waste, soil–emulsion mixes, and chemical additives, in partial or total replacement of coarse and fine natural aggregates, in addition to the use of piled embankments, for the reinforcement of the natural subgrade in the presence of a very thick, soft soil layer in the foundation of the highway.

**Keywords:** highway, pavement, soft soil, expansive soil, Synthetic Coarse Aggregate of Calcined Clay (SCACC), reinforced piled embankment, recycled material, Brazilian Amazonia

## 1. Introduction

In this chapter, the possible Engineering solutions for the problems faced in the implementation of road infrastructure in the Brazilian Amazon are discussed, seeking technical alternatives allied to the reduction of logistics costs and the reduction of environmental impacts.

## 2. Methodology

This study fundamentally addresses the general characterization of natural, technical, economic, and environmental problems for road construction in the Amazon environment, pointing out possible solutions to circumvent these difficulties. The characterization was performed by reviewing the available technical literature, while the solutions were indicated by case studies of works conducted in the region.

It should be noted, however, that while there is a long tradition of environmental studies in the Amazon, involving biotic and abiotic environments, and the environmental impacts resulting from anthropic actions, the same tradition does not apply to studies of road constructions, from the point of view of Engineering.

The main cause of this problem is the low rate of research data publication and its application on highways in the specialized technical literature.

Thus, although the application of a certain material or technique in each construction of a highway is known, it is difficult to make a deeper analysis, due to the absence of published data.

For this reason, the literature review and case studies discussed here were mostly restricted to academic research of master and doctoral thesis, presenting the results of field monitoring reported by their respective authors.

## 3. Case studies – results and discussion

In the preceding items, the general characteristics of the International Amazon were contextualized, with emphasis on the Brazilian Amazonia, in addition to addressing the problems faced in the construction of highways in this region. This Item presents some case studies that sought to circumvent the problems of scarcity of stone material near the construction site and the presence of soil layers with low bearing capacity in the natural foundation (subgrade) of the highways built in the Brazilian Amazonia environment.

**Figure 1** shows the locations of the case studies analyzed here.

### 3.1 Stabilization of Amazonian soils with chemical additives

An unpaved rural road located at 53 km of the state highway AM-010, north of the city of Manaus (Point 1 of **Figure 1**), is used to access a military base. This road has a natural subgrade consisting of a layer of yellow-red sandy silt containing a 55% silt fraction, 35% sand, and 9% clay and is classified as a latosol, with a 30% LL, a PI of 10%, a CBR of 31% and an optimal moisture content of 12.6%, both in the intermediate compaction energy. The soil pH is acidic, with a value of 5.39, kaolinite being the predominant clay mineral. The soil is classified as CL (USCS), A-4 (TRB), and LG′ (MCT).

The efficiency of chemically stabilizing the subgrade soil with cement supplemented with ZS was evaluated, resulting as a single layer of the stabilized base; for this purpose, was constructed an experimental section of 70 m in length and 7.5 m in width. In order to control the performance of executive technical procedures, SCS and TSDC tests were made before, during, and after construction, in addition to evaluating the overall quality of the road paving and its superficial drainage elements.

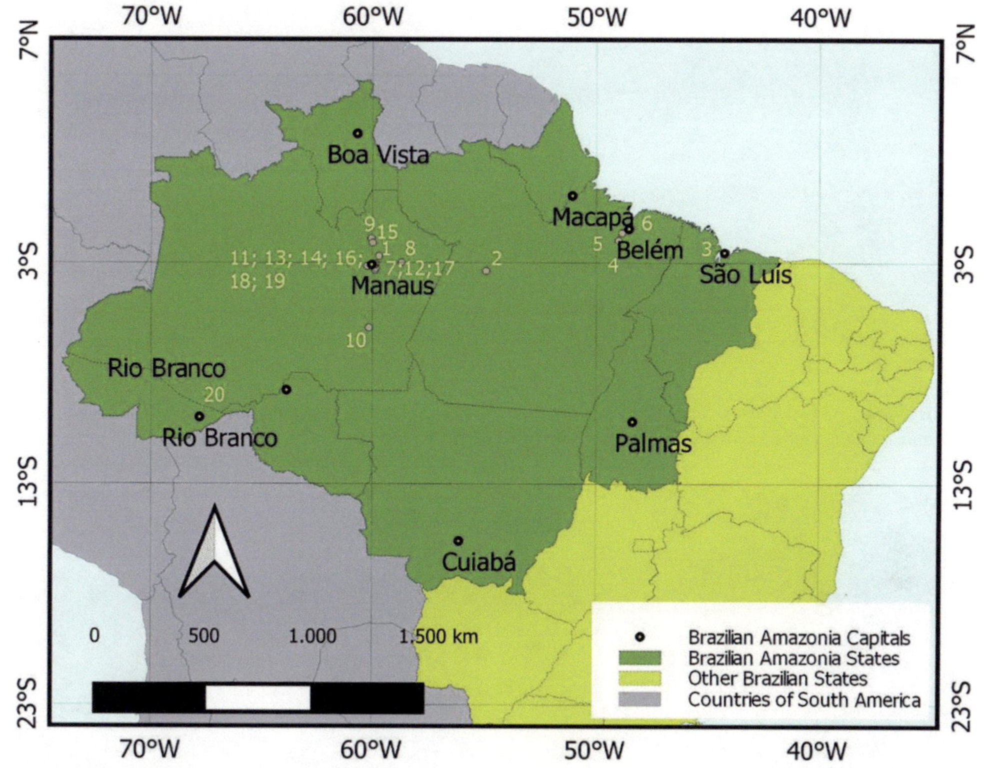


**Figure 1.**
*Locations of the case studies analyzed.*

The ZS-based additive (RoadCem®) used is a fine, odorless, grayish powder with a specific mass of approximately 1,100 kg/m$^3$ and pH of 10–12 (in water, at 20°C). Its chemical composition is mainly composed of alkali metals and alkaline earth metals (60–80%), including sodium, potassium, calcium, and magnesium chlorides; ZS and oxides (5–10%); and activators (5–10%).

Based on the characteristics of the experimental section subgrade soil, a soil-cement–the additive mix was made with 1.7 kg/m$^3$ of ZS (0.09% per dry mass) and 160 kg/m$^3$ of cement (8.20% per dry mass).

The SCS test results of the specimens molded in the laboratory before the fieldwork found values of approximately 8 MPa at 28 days. In general, the construction stages begin with the addition and mixing of the ZS additive and cement, followed by leveling, compaction, and surface finishing. In this process, the ZS additive may be added one day before the application of the cement (**Figures 2** and **3**).

The mechanical strength of the stabilized base was verified by molding PCs in three stages. In the laboratory, before starting the stabilization process, specimens were prepared with the pre-established dosage, and the SCS and TSDC test results were evaluated at the rupture ages of 3, 7, 14, and 28 days in wet curing.

During the execution of the stabilization step, a new molding of specimens was performed by collecting the homogenized mixture of soil-cement–ZS immediately before starting the field compaction procedure. In the third stage, after 28 days of stabilization, six samples were extracted directly from the runway. The laboratory

**Figure 2.**
*a) Distribution of cement and ZS additive in the experimental runway by manual spreading; b) homogenization of the soil, cement, and ZS additive in situ by means of a recycler [1].*

**Figure 3.**
*Steps of a) compaction; b) bulging; c) surface finishing; and d) curing of the stabilized base [1].*

results of the samples molded in the laboratory, those molded with a field mixture, and those extracted from the runway are shown in **Figure 4**.

Of the specimens molded with the field mixture during the construction, the SCS results were observed to be close to the values of those specimens molded in the laboratory. The quality of the mixture made by the recycler and the time between homogenization and compaction are the main factors underlying the difference between the SCS of the samples molded in the laboratory and molded with the filed mixture, around 16%.

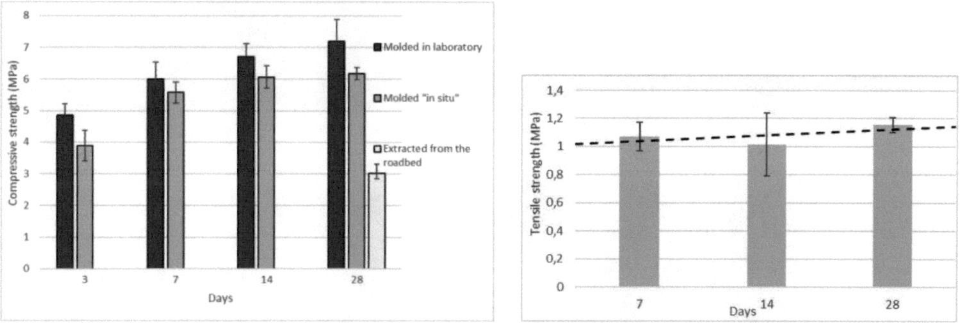


(a) (b)

**Figure 4.**
*a) Comparison of SCS for samples molded in the laboratory, during stabilization, and extracted directly from the experimental runway; b) TSDC for laboratory-molded samples [1].*

Conversely, the samples extracted directly from the runway reached only 42% of the SCS resistance value of the samples molded in the laboratory at 7 days of curing (**Figure 4a**), but these values were still higher than the minimum Brazilian road standard, which is 2.1 MPa. The difference is mainly caused by the construction techniques, the quality of the mixture, and compaction by the equipment in the field, under conditions not always optimal, while in the laboratory, technological control is easier to achieve. In addition, the extraction process of field samples can damage the cemented structure of the specimens, reducing their real strength.

Regarding the tensile stresses that the lower regions of the pavement receive due to the transient loads imposed by the traffic, the TSDC of the stabilized base material reached values above 1 MPa at a curing age of 7 days (**Figure 4b**).

In the visual inspection of the work during the first 48 hours of curing, some transverse, superficial, and isolated cracks with average depths of 1 mm emerged. In this period, there was the volumetric expansion that generated internal tensile stresses in the material that was still in the fresh state. The microstructure in the formation is subjected to secondary growth of the impure phase of ettringite; because of the continued hydration of $C_3S$, hydrated calcium silicate began to form inside the hydrated carapace [2]. This effect was more intense in plastic clays with high levels of cement exposed to the high humidity of the night-time forest and the intense heat of the day. However, the continuous strength gain of the material interrupted the propagation of cracks in later days, which naturally incorporated dust, showing an aspect of natural base regeneration. Four years after construction, the pavement showed no cracks, abatements, or pathologies that would compromise the durability and assimilation capacity of traffic surcharges (**Figure 5**).

### 3.2 Granulometric stabilization of Amazonian soils using SCACC

Cabral et al. [3] and Cabral [4] conducted laboratory and field studies on the use of SCACC in asphalt mixtures, with raw material from the state of Pará. The authors obtained excellent results when comparing SCACC to the natural coarse aggregate (crushed stone), indicating that the synthetic aggregate supports severe mechanical compaction, according to the results obtained in the degradation tests (analysis of the comparison between the granulometric composition of the SCACC in the conditions of no compaction on the experimental runway, and after two years of service

**Figure 5.**
*Final aspect of the experimental section. a) One day of stabilization, in Oct. 2016; b) after 490 days, in Feb. 2018; c) after 4 years, Oct. 2020 [1].*

**Figure 6.**
*Use of SCACC in highway paving: a) burnt of ceramic bricks following a controlled firing temperature; b) crushing of the ceramic blricks. c) Stacking of the mixture of lateritic soil and SCACC in the field [4].*

completion). The study site was an experimental section of the pavement restoration work on federal highway BR-163, between 101 km and 102 km (Point 2 of **Figure 1**), whose differential treatment was the incorporation of SCACC in the base layer of the pavement.

The base course of this segment was built in mid-November 2007, totalling 1,000 m in length, 12 m in width (two lanes of 3.5 m each and two shoulder lanes of 2.5 m each), and 20 cm thick. Not only in this section, but throughout the extent of the pavement restoration work, the project established the need to incorporate approximately 30% of coarse aggregate into the lateritic soil (*piçarra*) obtained from the deposits prospected by the geological-geotechnical studies.

**Figures 6** and **7** show the stages of the use of SCACC, from the preparation of the ceramic bricks (which were later crushed into coarse aggregate sizes) to the execution of the base course in the field.

### 3.3 Chemical stabilization of Amazonian soils with soil: Asphalt emulsions

To observe the behavior of an application of soil–asphalt emulsion mixture, Sant'Ana [5] studied a 200-m unpaved experimental section located at gate 3 of the access to the State University of Maranhão (Point 3 of **Figure 1**), which had a highly irregular surface, composed of a base course in laterite that was very deteriorated (**Figure 8**) overlying a subgrade layer formed by fine sandy soil (no subbase layer was observed). After the collection of deformed samples to characterize the existing

(a) (b) (c)

**Figure 7.**
*Use of SCACC in highway paving: a) scattering of the mixture of lateritic soil and SCACC for the execution of the base course; b) completed experimental runway, with asphalt course in double surface treatment. c) Conditions of the experimental runway approximately 2 years after completion [4].*

**Figure 8.**
*View of the experimental section before stabilization with asphalt emulsion [5].*

layers, satisfactory results of the subgrade were observed (CBR of 12% under the normal compaction energy). On the other hand, the material of the base layer fit the subbase properties (CBR of 23% under intermediate compaction energy), so it needed to be reinforced and regularized with material imported from a deposit, resulting in a new base course.

Regarding the stabilization of the new base layer with RL-1C asphalt emulsion (cationic, slow rupture), the surface (base) was regularized and the existing road widened, keeping the transverse slopes at 3–5% to allow the accelerated flow of rainwater. Then, over the first 100 m, scarification was performed at a thickness of 5 cm (**Figure 9a**), followed by wetting until the layer reached the optimum moisture content; homogenization with a disc harrow (**Figure 9b**); application of RL-1C asphalt emulsion at the rate of 5 $l/m^2$, corresponding to 5% by weight of the dry soil, but in three stages (**Figure 9c** and **d**); and homogenization with the disc harrow between asphalt emulsion steps. A grader was used to move the soil–asphalt emulsion mixture to the side of the road, sometimes on one bank, sometimes on another (**Figure 9e**). Next, the mixture was compacted with a smooth roller (**Figure 9f**). Finally, the preliminary sealing layer was applied, with the emulsifying

**Figure 9.**
*Some records of the execution stages of the experimental section with soil–emulsion [5].*

agent spreading (**Figure 9g**) and the release of clean sand (**Figure 9h**), which was then compacted again by a smooth roller. **Figure 9i** shows the finishing appearance of the stabilized base 7 days after execution. The remaining 100 m were subjected to the same procedure as above.

To evaluate the stabilization efficiency, two non-destructive tests using Falling Weight Deflectometer (FWD) equipment were carried out on the finished surface, one 12 days and the other 20 months after execution. The current service value of the road surface was characterized using the method developed by the United States Department of Army in 1995, with the objective of calculating the Unsurfaced Road Condition Index (URCI). Surface distress surveys were carried out in December 2007 and September 2008, that is, 10 months and 20 months after the stabilization was completed, respectively.

Although the material was in good general condition, the defects that were most common in the analyzed sections were surface dust, the loss of aggregate due to abrasion, and the "pothole" or "potholes". The classification of all sections of the experimental stretch as "very good" or "excellent" by URCI method, after almost 2 years of operation (**Figure 10**), showed that the soil–asphalt emulsion technique is applicable in the region where it was studied for roads with low traffic volume.

### 3.4 Reinforced piled embankment as a Foundation of Roads on soft soils in the Brazilian Amazonia

There are four cases to be reported here (Points 4 to 7 of **Figure 1**). All cases dealt with the construction of a road in a location with a thick layer of soil with low bearing capacity, in which it would be uneconomical to purge this material.

**Figure 10.**
*Surface condition of the stabilized base with asphalt emulsion after 20 months of execution [5].*

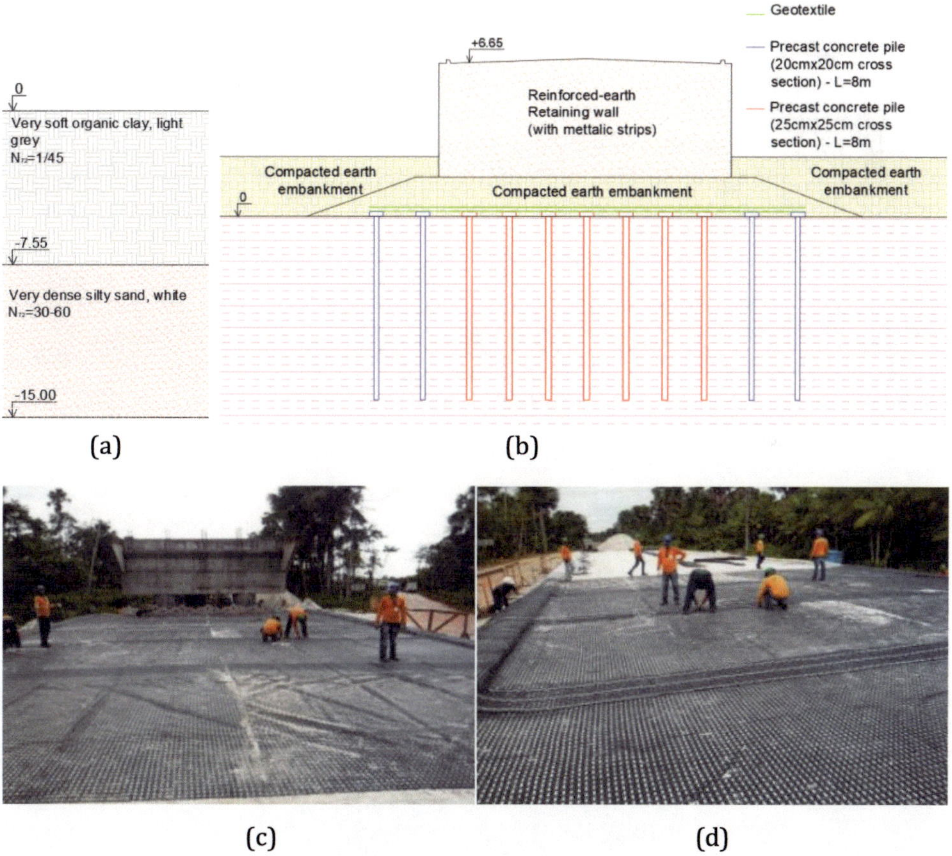


**Figure 11.**
*Reinforced piled embankment executed for crossing watercourses. a) Stratigraphic profile; b) arrangement of the piles and the geogrid reinforcement, in cross-section; c) and d) installation of the geogrid on the capitals of the piles (photo: Maccaferri [6]).*

**Figure 11** shows the implementation details of the project (Point 4 of **Figure 1**). The largest thickness of the very soft organic clay layer observed was 7.55 m (**Figure 11a**), with a penetration resistance index ($N_{SPT}$) value under net energy of 72% ($N_{72}$, typical of Brazilian energy test) with an average of 1/45 blows/30 cm, on which a road embankment and bridge abutments (with a height of 6.65 m) were built to cross a watercourse called "Igarapé Miri" in the State of Pará. **Figure 11b** indicates the solution adopted. It entailed precast reinforced concrete piles, with square sides of 20 cm (to support the embankment) and 25 cm (to support the embankment and the bridge abutments), that were 8 m in length, spaced every 1.5 m, in both directions in plan view. On average, the pile tips were embedded at least 1 m into the resistant layer of very dense silty sand. **Figure 11c** and **d** shows a high stiffness geogrid for reinforcement of the embankment, installed over the cap piles, with the function of distributing the loads of the embankment and the bridge abutments for the piles.

**Figure 12** shows the implementation details of the project (Point 5 of **Figure 1**). The largest thickness of the very soft organic clay layer was 13.30 m (**Figure 12a**), with a mean $N_{72}$ value of 1/70 blows/30 cm (region under the tidal influence, located in mangrove), on which a road embankment and bridge abutments (with a maximum height of 4 m) were built to cross a small watercourse in the city of Abaetetetuba, Pará State.

**Figure 12b** indicates the solution adopted: There were used precast piles of reinforced concrete, with square sections of 35 cm on the side, 20 m in length, spaced every 3 m in plan view. On average, the tips of the piles were embedded approximately 5 m inside the resistant layer of very dense sand. A geogrid was used to reinforce the embankment, and a geotextile of high stiffness was placed between the geogrid and the top of the pile cap to avoid tearing of the geogrid when it was subjected to the

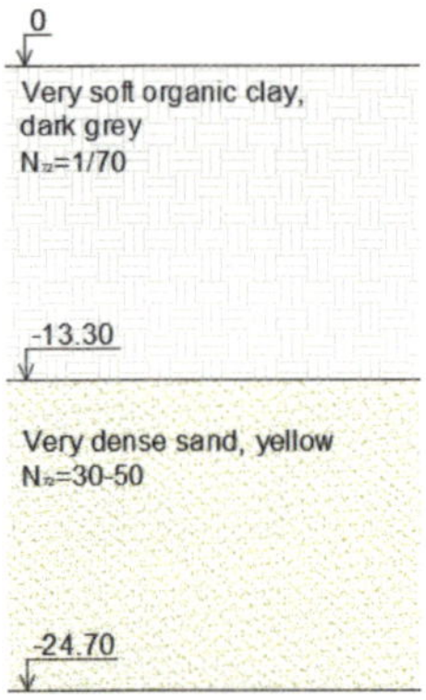


(a)

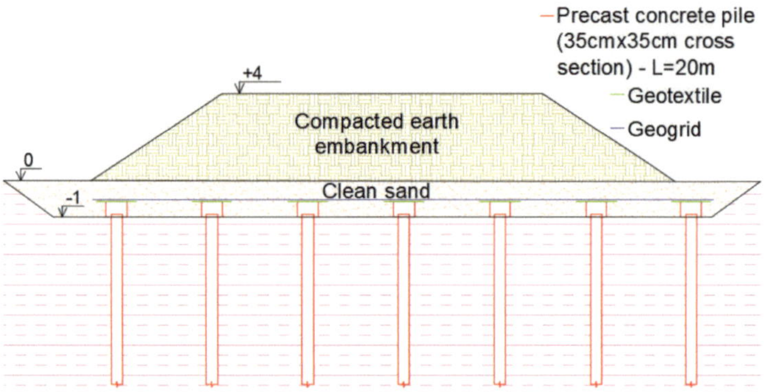


(b)

(c)

**Figure 12.**
*Reinforced piled embankment built for crossing the watercourse in the city of Abaetetetuba, Pará state. a) Stratigraphic profile; b) arrangement of the piles and the geogrid reinforcement, in cross-section; c) installation of piles in the abutment bridge (photo: NS Campelo).*

embankment surcharge. **Figure 12c** shows the piles supporting the retaining wall of the bridge abutments.

**Figure 13** shows the details of the project (Point 6 of **Figure 1**), a connection road to state highway PA-150, in the State of Pará. The largest thickness of the very soft organic clay layer observed was 8.00 m (**Figure 13a**), with a mean $N_{72}$ value of 1/45 blows/30 cm, on which a road embankment and an abutment bridge were built with a height of 2.3 m. **Figure 13b** indicates the adopted solution. There were used precast reinforced concrete piles, with 25 cm square side, spaced every 1.5 m. On average, the tips of the piles were embedded approximately 1 m inside the resistant layer of very hard silty clay. A geogrid and a geotextile were placed at the top of the pile caps.

**Figure 14** shows the implementation details of the project (Point 7 of **Figure 1**). The highest thickness of the soft clay silt layer observed was 5.75 m (**Figure 14a**), with a mean $N_{72}$ value of 3 blows/30 cm, on which a road embankment with a height of 4–6 m was built, in the *várzea* region, on the BR-319 federal highway, State of Amazonas. **Figure 14b** indicates the solution adopted: root-type piles 40 cm in diameter, spaced every 2 m. A geogrid was used to reinforce the embankment, and a geotextile mat of high rigidity was placed between the geogrid and the top of the pile capitals to avoid tearing the geogrid.

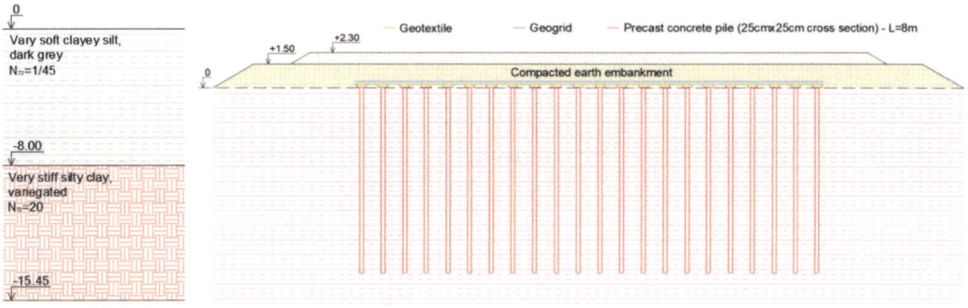


(a) (b)

**Figure 13.**
*Connection road to state highway PA-150, state of Pará. a) Stratigraphic profile; b) arrangement of the piles and the geogrid reinforcement, in cross-section.*

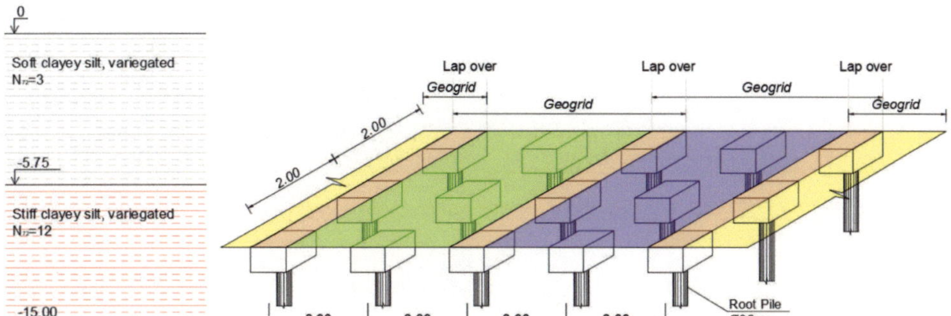


(a) (b)

**Figure 14.**
*Reinforced piled embankment executed in the várzea region on the BR-319 federal highway, state of Amazonas. a) Stratigraphic profile; b) arrangement of the piles and the geogrid reinforcement, in cross-section.*

### 3.5 Use of lateritic concretions in road pavements in the Amazon

As reported in Item 2.8 of the Chapter "Challenges in the Construction of Highways in the Brazilian Amazonia Environment: Part I: Identification of Engineering Problems", laterites are widely used in road pavings in the Brazilian Amazonia, but their occurrence is limited in some highland regions (*terra firme*), where the conditions are favorable for their formation. In addition, the deposits are not very thick, so the available volumes are not high. Thus, it is a "noble" material, and its use in highways is limited to the base course and in primary coatings (protection of the lower layers until the execution of asphalt course) of pavements with high traffic volume (**Figure 15a** and **b**) or only as a primary coating on rural roads with low traffic volume (**Figure 15c** and **d**). **Figure 15b** and **d** refers to Points 8 (state highway AM-363, "Estrada da Várzea") and 9 (Ramal Sargento Picanço, 123 km of federal highway BR-174) of **Figure 1**, respectively. Note the different shades of the lateritic materials.

**Figure 16a** shows some detail of a typical coarse lateritic concretion (*piçarra*) of the Amazonia, with a matrix in iron and aluminum oxides. There are records of roads with decades of construction which were used primary coatings of lateritic concretion, as shown in **Figure 16b** (Point 10 of **Figure 1**), on state highway AM-174, and even without the existence of a rainwater drainage system, it still shows satisfactory trafficability almost the entire length of the road.

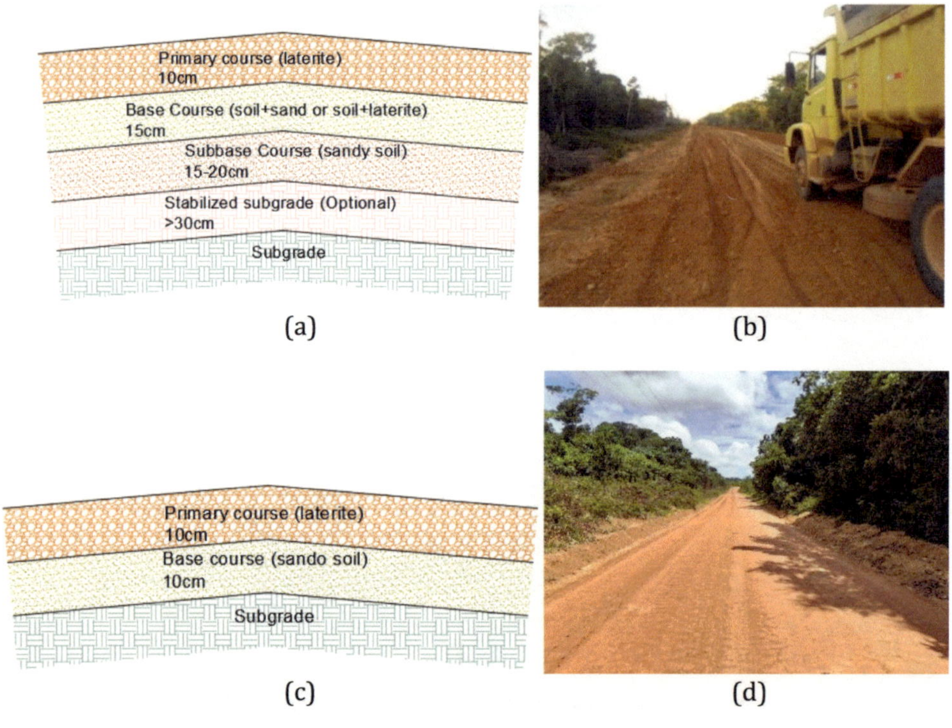


**Figure 15.**
*Use of piçarra on highways. (a) Typical profile of the pavement courses for high traffic volume; (b) execution of the primary coating layer; (c) typical profile of the pavement layers for low traffic volume; (d) condition of the primary coating layer performed over decades (photos: NS Campelo).*

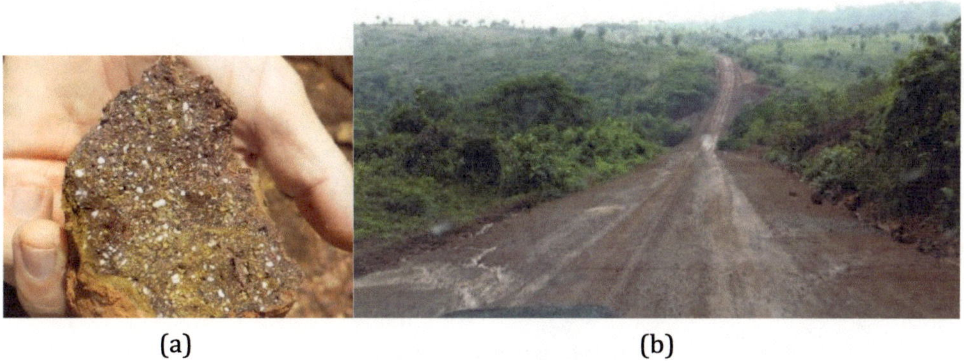

**Figure 16.**
*Use of laterites in Amazonian highways. a) Detail of lateritic concretion, with a predominance of iron and aluminum oxides; b) primary coating layer aiming at the durability of the pavement until the execution of the asphalt course (photos: NS Campelo).*

### 3.6 Reinforcement of the natural Foundation of the Highway by partial soil exchange and replacement by a drainage structural layer

Sometimes, when the layer of low bearing capacity of the natural foundation (subgrade) is not very thick (2–3 m), then it is possible to replace all of it with another, better-quality material, imported from the deposit. When the layer is deeper (**Figure 17a**), and the total removal of the layer or the use of reinforced piled embankment is uneconomical (discussed in Item 4.4 of the Chapter "Challenges in the Construction of Highways in the Brazilian Amazonia Environment: Part I: Identification of Engineering Problems"), it is possible to partially replace the layer by introducing a layer of coarse crushed stone material (locally known as *rachão*, with a particle size of 100–250 mm, i.e., cobble grain size classification) over a layer of clean sand (**Figure 17b**). The first layer acts as a structural reinforcement, while the second layer acts as a draining layer, although both layers are very permeable and provide a significant increase in the shear strength of the natural foundation layer. Sometimes, geotextile is placed between the natural subgrade and the sand layers, and between the layers of the *rachão* (high stiffness geotextile must be used so as not to be ripped by the friction with the crushed stones) and the compacted embankment, to avoid filling the voids of these layers. However, geotextile is not introduced between the sand and *rachão* layers, as it is intended that there be "needling" of the sand particles in the voids of the rock granular material, allowing the internal structure of the grains to become more rigid by particle-to-particle contact.

**Figure 18** shows the phases of partial excavation of the soft clay layer for the execution of an urban connection corridor in the city of Manaus, Amazonas (Point 11 of **Figure 1**). The thickness of the *rachão* layer – despite being a bearing capacity problem and being predefined in the design – is usually confirmed in the field by local experience in passing a fully loaded truck (82 kN of load for a rear single axle with dual wheels) several times over the surface of the layer, and no visible deformation is observable to the naked eye. In general, this thickness varies between 60 and 100 cm and may reach an even greater value, depending on the site conditions and embankment surcharge. Obviously, this is an onerous solution for locations where stone material is scarce and therefore is limited to short stretches of a road.

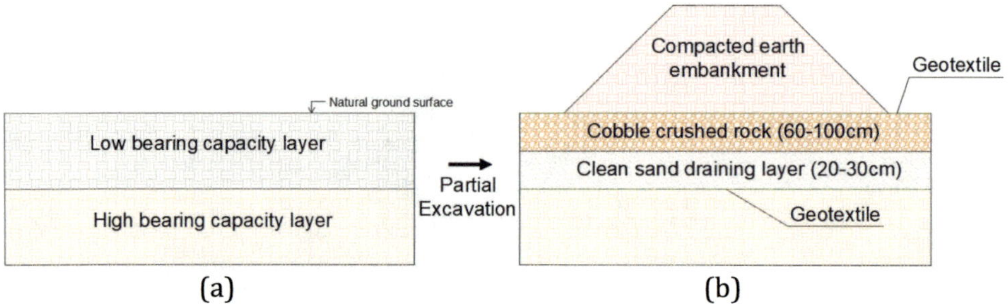


**Figure 17.**
*Partial replacement of natural foundations with low support capacity by drainage layers and greater shear strength in road construction.*

**Figure 18.**
*Construction of an urban corridor with partial replacement of the natural ground with low bearing capacity. a) Partial excavation of the layer; b) placement of the coarse crushed stone material (rachão); c) placement of the embankment layer for road construction; d) appearance of the rachão layer after partial implementation of the compacted earthen embankment (photos: NS Campelo).*

## 3.7 Construction of highways in the *Terra Firme* and *Várzea* regions

Item 1 described that the Amazon Basin encompasses a vast area of lowland and relatively flat lands, with numerous watercourses crossing them. The highlands, which are not flooded, constitute the *terras firmes*, while the lowlands areas flooded by flood river pulses constitute the *várzeas*. In the *terras firmes,* there is a predominance of yellow-red latosols, and horizons of *piçarras* are found with relative ease, albeit with reduced volume, given the low thickness of these layers. In addition, the subgrade in these regions usually has more permeable material and, together with the more rugged relief, allows better surface and deep drainage of the highways.

However, in *várzea* regions, there are several complications for highway construction. Starting with the impermeable natural subgrade (silty or clayey soil), formed mainly by argisols and gleysols, with low bearing capacity (mean $N_{72}$ less than 3 strokes/30 cm, which could meaning an allowable bearing capacity of soil somewhat less than 30 kPa, at depths ranging from 0 to 4 m in most cases), an absence of horizons of lateritic concretion, a deficiency of quality material to constitute the road embankment (generally, the existing soils consisting of expansive clay minerals, even without the presence of organic matter), an absence of rocky material, and a natural drainage basin forming a tangle of watercourses with approximately 10–15 m of vertical fluctuation between the maximum flood and minimum ebb levels, etc.

**Figure 19** shows two Brazilian federal highways that cross the State of Amazonas. **Figure 19a** shows a stretch of highway BR-174, with a total length of 3,320 km, which connects the States of Mato Grosso, Rondônia, Amazonas, and Roraima and from this to Venezuela and the rest of the Americas and Caribbean (see **Figure 7** of the Chapter "Challenges in the Construction of Highways in the Brazilian Amazonia Environment: Part I: Identification of Engineering Problems"); only 210 km of this highway cross the state of Amazonas. Throughout almost its whole length, deposits of *piçarras* and stone material are found; it cuts the *terras fimes*, with the top of the highway located between 50 and 200 m altitude; it crosses a more dispersed natural drainage basin, with lower frequency of larger watercourses. **Figure 19b** illustrates a stretch of highway BR-319, with a total length of 885 km, of which 820 km is in the state of Amazonas and 65 km is in the state of Rondônia (see **Figure 7** of the same Chapter), connecting the city of Manaus to the city of Porto Velho in the North–South direction and from there to the rest of the country. The highway crosses *várzeas* for much of its length (approximately half the length, in the direction Manaus–Porto Velho), with deposits of lateritic concretions and rocky material only in the last 200 km of the highway. The top of the highway is located at elevations between 25 and 70 m altitude and crosses a more concentrated natural drainage basin, with a higher frequency of small and large watercourses.

Due to the period of river flooding in the *várzea* region, the highways function as "earth dam", dividing the drainage basin between the two sides. In fact, the situation is more aggravated because, while in the conventional dams only the upstream side

(a) (b)

**Figure 19.**
*Different reliefs cut by highways in the state of Amazonas. a) Wavy, in the "terra firme" region. b) Flat, in a "várzea" region (photos: NS Campelo).*

is subject to the variation in the external water level, in these highways both sides are influenced by the flooding and ebbing water levels. The ebb period is the most critical because, as the embankment is formed mainly of clayey soils, the dissipation of pore-pressures from the interior of the massif is slow and, if accompanied by a rapid decrease in the water level of the surrounding watercourse, may result in a phenomenon analogous to the slope failure by "rapid drawdown", observed in conventional dams (**Figure 20**). Thus, any of the highway margins may be subject to slope failure, and there may even be a rupture of both sides. Therefore, there is a combination of factors that can lead to the rupture of these road slopes: large increase in the external water level; slow dissipation of pore-pressures from the interior of the clayey/silty compacted embankment; and natural subgrade formed by soil stratum of low to medium bearing capacity.

Large river floods have become more frequent in the Amazon Basin, leading to increasing maxima water levels year by year. Thus, in some regions of the BR-319 highway, overtopping occurs (**Figure 21**), and in the ebb cycle, partial rupture of the slopes may occur in the higher embankments when conditions are favorable (**Figures 22** and **23**).

**Figure 22a** (point 12 of **Figure 1**) is located near 23 km of highway BR-319, while the other occurrences are in a range between 20 km and 60 km.

Souza [7] and Souza et al. [8] studied the phenomenon of *terras caídas* ("fallen lands"), which is the rupture of riverbanks, commonly in floodplain regions, under the ebb of "white-water" rivers (classification due to Sioli [9]), locally called "águas barrentas" ("muddy waters"). Despite being a natural phenomenon, the same

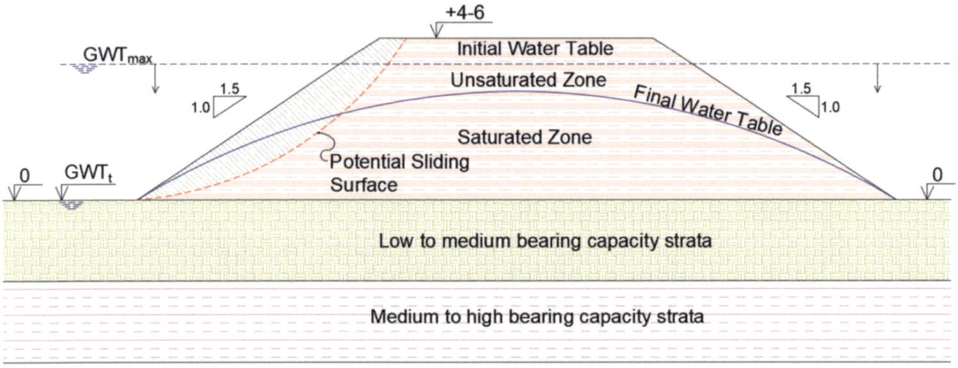


**Figure 20.**
*Lowering of the river water level on both sides of the road.*

**Figure 21.**
*River flooding causing overtopping at some places on federal highway BR-319 (photos: NS Campelo).*

**Figure 22.**
*Rupture of road slopes, after the river ebb cycle (photos: NS Campelo).*

**Figure 23.**
*Rupture of road slopes, after the river ebb cycle (photos: NS Campelo).*

concept of hydraulic rupture can be applied to road embankments, as described in **Figure 20**. **Figure 24** shows the phenomenon of "fallen lands" near the city of Manaus (Point 13 of **Figure 1**).

This phenomenon causes damage to the riverine population, as a large amount of land detaches from the massif, reaching the residences (wooden *palafitas*) and the urban and rural roads that border the riverbank of the communities and cities. It can also cause small "tsunamis" when a rupture of land occurs, causing the sinking of small boats anchored nearby.

### 3.8 Behavior of tropical soils under suction pressure and the laterization process

As reported in Items 2.8 and 2.9 of the Chapter "Challenges in the Construction of Highways in the Brazilian Amazonia Environment: Part I: Identification of Engineering Problems", tropical soils present some behaviors dictated by the pedogenetic process of laterization and soil suction when in an unsaturated state. Together, they significantly increase the resistance of cutting (mainly) and embankment slopes (in relation to the matric suction) in the case of shear resistance against rupture and erodibility processes.

There are records of cutting slopes existing for more than 50 years that are almost vertical (**Figure 25a** and **b**), which, even without anti-erosion protection – by vegetation cover (grasses or native vegetation) – or even without the existence of any superficial drainage of rainwater (in a region of high rainfall, above 2,500 mm per year, see **Figure 2b** of that Chapter), remains in a stable condition, except for some non-lateritic stratum (saprolitic), where it is vulnerable to erosion (**Figure 25b** and **c**), or when it is sandy or silty lateritic soil (**Figure 25a**). **Figure 25a, b**, and **c** corresponds to Points 14 to 16 of **Figure 1**, respectively. Other times, the distinction between the lateritic and saprolitic horizons is given by the existence of a "stone line" of lateritic concretions, as shown in **Figure 26a**, on state highway AM-070 (Point 17 of **Figure 1**).

In some cases, a thin film (1–3 mm) of lateritic crust forms on the surface of the slope, of dark red color, which acts as an anti-erosion protectant, allowing the stability of the slope against the deleterious action of the rains for decades, even with almost vertical slopes. **Figure 26b** and **c** shows these thin films (Points 18 and 19 of **Figure 1**, respectively) that occur in the city of Manaus.

### 3.9 Ceramic, recycled, and alternative materials in road paving of Amazonian highways

There are reports of the use of milled asphalt waste (reclaimed asphalt pavement-RAP) from the recycling of asphalt coatings on federal highways BR-163 and BR-364 in the State of Mato Grosso [10] but in the initial state. An interesting application

**Figure 24.**
*Natural phenomenon known as terras caídas ("fallen lands") occurring in the Careiro da Várzea region, in front of the city of Manaus (photos: NS Campelo).*

(a) (b) (c)

**Figure 25.**
*Cutting slopes. a) Erodible horizons (sandy/silty soil). b) and c) lateritic and saprolitic horizons (highly erodible) (photos: NS Campelo).*

of ceramic materials occurs in some urban streets in the city of Rio Branco, State of Acre, where traditional ceramic bricks are applied as linings in streets with low traffic volume (**Figure 27** – Point 20 of **Figure 1**).

There are no reports of the use of lightweight fill materials (e.g., expanded polystyrene – EPS) for embankments to be constructed over soft soils in road paving in the Brazilian Amazonia. This type of solution could be used in highways that cross natural subgrade in *várzea* terrains.

In general, there is a lack of technical information on the construction procedures and use of construction materials on the highways of the Brazilian Amazonia, which is reflected in the few publications on the subject.

**Figure 26.**
*a) Presence of a "stone line" (lateritic concretions) separating the lateritic and saprolitic horizons. b) and c) presence of thin film of lateritic crust in slopes (photos: NS Campelo).*

**Figure 27.**
*Use of ceramic block as a coating for urban roads with low traffic volume (photo: Jornal AcreAgora (https://acreagora.com/2021/07/12/).*

## 4. Conclusions

In order to overcome technical problems reported in the Chapter "Challenges in the construction of highways in the Brazilian Amazonia environment: Part I: Identification of Engineering Problems", can be employed solutions with materials found in the vicinity, in most cases. Thus, for problems of the absence of stone material at the construction site, the use of natural materials (lateritic concretion), synthetic materials (SCACC), and recycled materials (e.g., from ceramic waste and milled RAP) can be used, while for problems of the natural foundations of roads with low bearing capacity can be used partially or fully replacement of natural soil for a layer composed of coarse crushed stone (*rachão*) and clean sand, or, for greater depths, it can be used reinforced piled embankment (fill supported on piles and reinforced by geogrids).

Although the Brazilian Amazonia brings some hindrances to the construction of highways in the *várzea* regions, nevertheless in the elevated regions (*terra firme*), lateritic concretions can be counted on for use in base and primary courses, and there are lateritic soils of better quality for the construction of embankments, in addition to the favorable conditions of slope stability, especially in periods of drought rain, induced by soil suction.

For rural roads with low traffic volume, ceramic waste can also be used as recycled material in primary coatings.

## Acknowledgements

The authors would like to thank CAPES for the grant of a scholarship for an academic master's degree to the Civil Engineer Daniel Jardim Almeida.

## Author details

Nilton de Souza Campelo[1*], Arlene Maria Lamêgo da Silva Campos[1],
Marcos Valério Mendonça Baia[1], Daniel Jardim Almeida[1],
Raimundo Humberto Cavalcante Lima[2], Danielly Kelly dos Reis Dias[1],
Júlio Augusto de Alencar Júnior[2] and Mário Jorge Gonçalves Santoro Filho[3]

1 Graduate Program in Civil Engineering (PPGEC), Federal University of Amazonas (UFAM), Manaus, AM, Brazil

2 Graduate Program in Geosciences (PPGGEO), Federal University of Amazonas (UFAM), Manaus, AM, Brazil

3 Department of Civil Engineering (DEC), Federal University of Pará (UFPA), Belém, PA, Brazil

4 STP Engineering Projects & Constructions, Manaus, AM, Brazil

*Address all correspondence to: ncampelo@ufam.edu.br

## References

[1] Almeida DJ. Stabilization of Tropical Clayey Soil with Synthetic Zeolite Cement as a Paving Solution in Amazonas. Manaus: Federal University of Amazonas; 2018

[2] Scrivener KL. The microstructure of concrete. In: Skalny JP, editor. Materials Science of Concrete I. Columbus: American Ceramic Society; 1989. pp. 127-163

[3] Cabral GLL, Motta LMG, Lopes LAS, Vieira A. Calcined clay aggregate: A feasible alternative for Brazilian road construction. In: Ellis E, Yu HS, McDowell G, Dawson A, Thom N, editors. Advances in Transportation Geotechnics. London: Taylor & Francis Group; 2008. pp. 90-96

[4] Cabral. GLL Pavement works and technology improvement [thesis]. Rio de Janeiro: Federal University of Rio de Janeiro (UFRJ/COPPE); 2011 (in Portuguese)

[5] Sant'Ana WC. Contribution to the study of soil emulsion in pavements of low-traffic highways for the State of Maranhão [thesis]. São Paulo: University of São Paulo (USP), Brazil; 2009 (in Portuguese)

[6] Maccaferri. Case history: Ponte Igarapé-Miri. Brazil: Pará State; 2016. p. 2 (in Portuguese)

[7] Souza ES. Fallen Lands ("Terras Caídas"): Characterization and geotechnical modeling of the Amazonian erosive phenomenon [thesis]. Manaus: Federal University of Amazonas; 2019 (in Portuguese)

[8] Souza ES, Campelo NS, Lima RHC, Aguiar RL. Geotechnical characterization and modelling of the "Fallen Lands" phenomenon in the amazon environment. Global Journal of Engineering and Technology Advances (GJETA). 2021;**9**(3):122-132. DOI: 10.30574/gjeta.2021.9.3.0168

[9] Sioli H. Das Wasser in Amazonasgebiet. Forsh. u. Fortschr. 1950;**26**(21-22):274-280

[10] Rota do Oeste. News; 2021 (in Portuguese)

Chapter 7

# Development of Highways Management Systems in Iraq

*Abdul Ridha Mohammed Afrawee, Haider Habeeb Aodah and Hussein Ali Mohammed*

## Abstract

The highways in Iraq need to be reviewed and evaluated urgently because of the poor condition of the roads, the lack of maintenance and the lack of services. This study provides specific criteria for assessing the condition of the road and determining the level at which it needs to intervene and perform maintenance. The study also proposes a new system for road management in Iraq, which contributes to providing the necessary funds for the continuity of maintenance work and the provision of services that contribute to the convenience of road users. Six criteria were determined to evaluate the efficiency of the road by conducting a survey of the opinions of engineers and academics specialized in the field of roads and these criteria are (Safety, Road Condition, Geometric Design, Cost, Services and Environment) and using the Statistical Product and Service Solutions Program (SPSS) a model was found that links road quality with these criteria. The Toll Road System (TRS) was proposed to collect tolls from road users. This system was compared with the current system in road management in Iraq using the Analytical Hierarchy Process (AHP) program, and the TRS was considered satisfactory by 75% of the sample surveyed.

**Keywords:** highways management system, highway data, criteria, SPSS, AHP, highway quality, TRS

## 1. Introduction

During the period of roads use, there are exposed to many natural, consumer and environmental factors that lead to their erosion and the appearance of various deformations in them and defects that cause economic damage and the safety and comfort of travelers on the road. Therefore, maintenance must be carried out periodically and not be delayed, because delaying maintenance causes an increase in the costs required for repairs, and maintenance is of two types, the first is periodic maintenance, and it must be carried out regularly on a daily basis to remove any waste that causes damage on the roads, and the second is emergency maintenance that is performed in cases of various accidents. These procedures require a fixed and stable financial allocation sufficient to maintain the quality of the road and the services provided on it.

Iraq has suffered from a lack of infrastructure and a low grade of development; the transportation sector is no exception. Reasons such as the absence of sectorial national and local strategy, chronic underfunding, poor institutions, and the fact that the environment is complex and influenced by the ongoing conflict have all contributed to making this situation even more difficult. To garneted fundamental needs, infrastructure investment is necessary. The need for efficient organizations to convert these investments into successful initiatives grows as well. However, beyond the problem of investment, there is a need for a fundamental sectorial change that is followed by the presence of a subsequent program that supports work on defining a strategy and vision while also working on building institutions and implementing investments [1].

This study establishes models for measuring the status of roads in Iraq based on essential criteria for determining road quality, allowing decision-makers to adopt the appropriate maintenance actions. It also provides a new method for road management that contributes to finding the necessary and continuous funding for the implementation of maintenance and development works for roads.

Highway management has significant implications as the management of public infrastructure is directly related to economic and social well-being and sustainability. Given the significance, the subject of road management in the field of public budgeting and finance has not been highlighted in comparison with other budgeting issues [2].

Highway network construction entails significant and permanent investments, necessitating thorough simulations and analysis before proceeding with the implementation choice. Because of changes in political, social, and environmental circumstances, this decision-making procedure is fraught with uncertainty. The concept also includes a multi-stage stochastic model for highway development decision-making, operation, development, maintenance, and rehabilitation [3].

In recent years, the highway network system has evolved at countries in terms of distances, scale, complexity, and technical performance. A great deal of scientific research has been done, and the results have substantially simplified the process of highway construction and development capacity. This began with the historical expansion of the road and highway network, which was based on certain construction and operation features. It starts with an orderly study of the important achievements in terms of construction techniques, operation, and control, and then moves on to the repairing and maintenance [4].

Road and highway traffic management is a sector of logistics which interests in the planning and controlling of traffic from one location to another. The major duty of traffic management is how to circulate traffic in a secure and effective style to minimize and eliminate accidents. Roads and highways mostly have devices prepared for traffic control, and most of them including live communication with the road and highway user, such as signals, signs and pavement's marks. These devices can help with navigation, determine the right-of-way, the places of safe passage, mark the speed limits, give instruction, warnings of hazards. Other types of traffic control measures include curbs stones, middle barriers and others [5].

Highways represent a key component of public property, and they are subject to a variety of deteriorations throughout the course of their lifespan, reducing their service life. Making judgments about maintenance, repair, and rehabilitation based on information about the current state, the hazard of its usage, prices, and the life cycle's age is an excellent duty for road and highway directors. Because road management is challenging, determining the best course of action for specified interventions is frequently left to the road managers [6].

With the speed of improving new information and communication technologies, there is a chance for arranging a total highway management system which could help in managing highway components in a quite effective and productive style. Worthy research and application efforts have been tested by highway officials in the past two decades in the field of road and highway management [7].

The transportation discipline has evolved significantly in response to the ever-increasing population. People's lifestyles are changing, and faster transit to different locations is required. As a result of the increased demand, new types of transportation such as air travel and rail and subway transit at or below ground level have emerged. While public transportation developed, the private transportation industry developed at a far faster pace, owing to features such as accessibility, appropriateness, flexibility, privacy, and luxury. This contributes to the increase in traffic, especially in the private transportation sector. As a result, the existing road space was insufficient to meet the increased traffic demand, and congestion developed. Furthermore, the accident's happening further enlarge [8].

Governments are often face serious challenges in financing huge infrastructure projects. Hence, governments usually depend upon imposing taxes on citizens. Private financing however has its own drawbacks. According to (Cruz and Sarmento 2017) [9], if the total evaluation for money value is needed for a project, the cost of capital is an important parameter that is affected by various kinds of determinants such as loan-specific, project specific, or economic environment-specific.

Many cities use toll road to relieve traffic congestion on commuter routes, often with the help of private enterprises. In exchange for toll income, private companies may build new roads or add lanes to existing highways. Cities manage traffic more effectively, toll operators improve incomes, provide better services, and keep roads in good condition. (Sanchez, December 1, 2015) [10].

According to (Jia and Tian 2002) [11], **Figure 1** depicts a flow chart for predicting a highway toll road. Adequate road operation should ensure the satisfaction of road user and quality of service provided.

Iraq had roughly 39,000 kilometers of paved roads and highways in 2005, the majority of which were massive roads constructed for commercial and military use in the 1970s and 1980s. After the 1991 War, when the infrastructure and road network were attacked, most deficiencies in highways and bridges were repaired. However, beginning in 2003, important roadways, such as the connecting route between Baghdad and the Jordanian border, were harmed more severely by security events. Bridges damaged in 2003 by coalition army was the center of major repair process in 2004 [12].

Despite attempts and expenditures to improve the serviceability of Iraq's roads and highways networks, the implementation and maintenance of these networks continues to deteriorate. The common occurrence and increase in the degree of deformity, as well as the lack of repeated maintenance, are the primary causes of such deterioration. Using the Pavement Condition Index (PCI) technique, many objectives were created to evaluate the pavement condition during visual inspection surveys. Several research have used Micro PAVER application software to assess PCI values based on GIS data. The findings indicated that certain roads were in terrible condition, while others were in good shape. It is critical to create a database for road problems, which should be updated regularly to account for the annual fluctuation in the PCI rate [13].

The State Commission of Road and Bridge (SCRB), One of the commissions under the Ministry of Construction and Housing (MOCH) is in charge of highways and road construction. In addition to maintenance and restoration, it generally entails supervision, labeling, and signage. Despite the fact that the SCRB possesses the necessary road

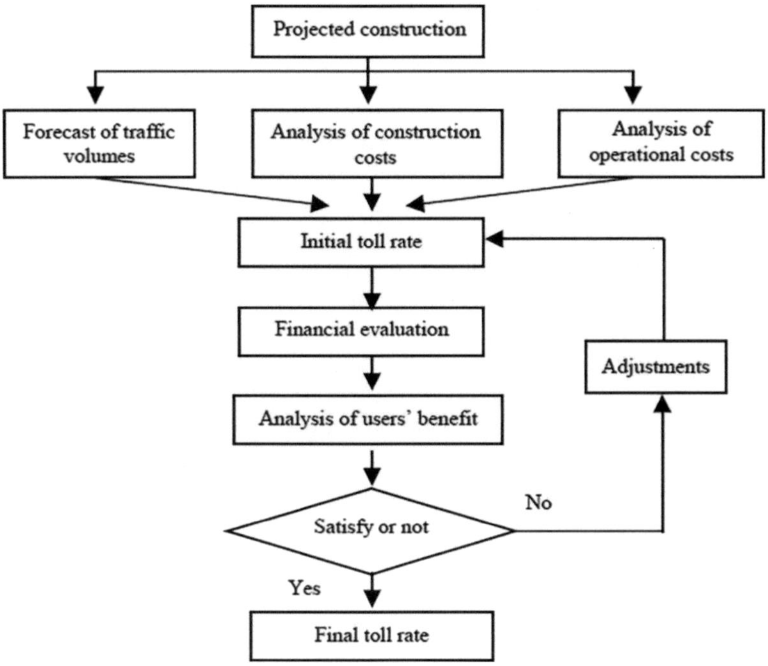


**Figure 1.**
*Procedure for toll rate determination.*

equipment to execute emergency work, the majority of road overlay work is normally done by contractors. Foreign construction companies are frequently hired to do the work. Private sector businesses are only allowed to participate in projects that need a high level of skill in road design and construction.

## 2. Aims of study

Some of Iraq's main cities are connected by a network of roads, which serve as the primary routes for the movement of people and products. It connects the country's different towns to the country's only sea port, which is located in Basra province and is considered the country's economic artery. After 2003, there was a massive rise in the number of automobiles, due to a change in the regime and increased purchasing power, as well as population growth. Highways, on the other hand, have not seen a rise in proportion to this expansion; in fact, they have suffered from a lack of maintenance. Highways in Iraq require effectual management systems for current restoration and maintenance, as well as enlargement and long-term sustainability. This research aims to provide an effective management system based on specialized databases and computer programs, guaranteeing that sufficient finances are available for highway maintenance and the establishment of standards.

## 3. Methodology

The study went through several stages for the purpose of completing the data and reaching the results. The first stage was to collect information and data from the concerned departments and previous documents and studies. The next stage was to conduct multiple questionnaires for different segments of road users, specialists and experts in the field of roads. After completing the data, computer programs were used. To reach the final results, the SPSS program was used to analyze the statistics and find road quality model, while the AHP program was used to compare the alternatives and choose the best ones.

## 4. Data analysis and results

This section deals with the surveys that were conducted with road users, road specialists and experts, and the results obtained from processing this data in the SPSS program.

### 4.1 The drivers' questionnaire

A questionnaire was distributed to road users and drivers of various types of vehicles to gauge their feelings about the current state of the road and their willingness to accept the implementation of the toll road system. The sample consisted of 94 users, whose responses are shown in **Table 1**. The table shows that 77% refuse to pay the fees for using the road in its current condition, while 87% agree to pay the fees if maintenance is done for the road and the provision of services and safety factors.

Also, 63% of the respondents indicated that they would increase the fare for passengers if the fees were collected from them to use the road. Finally, 94% of them expressed their desire to limit the speed on the expressway and impose fines on violators.

**Figure 2** depicts the order of services sought by road users in order to optimize efficiency.

### 4.2 Questionnaire for specialized academics

The questionnaire items included a group of carefully designed questions that aim at determining and describing the quality and operational performance of the road based on a set of main criteria from which a set of sub criteria were derived **Table 2**. From the table the highest importance was given to safety by 9 degrees, and the least importance in determining the efficiency of the road was given to the

| Article | Yes | No | % Weight of yes answer |
|---|---|---|---|
| Are you ready to pay fees for using the highway as it is? | 22 | 72 | 23.40 |
| If the highway is rehabilitated, maintained and furnished with lighting and traffic signs, provided with services and safety requirements, are you ready to pay the fees? | 82 | 12 | 87.23 |
| Would you increase the travel fare if you were charged a fee for using the highway? | 60 | 34 | 63.83 |
| Do you prefer speed limitation and fines for those who exceed it? | 89 | 5 | 94.68 |

**Table 1.**
*Percentage weights of questionnaire questions.*

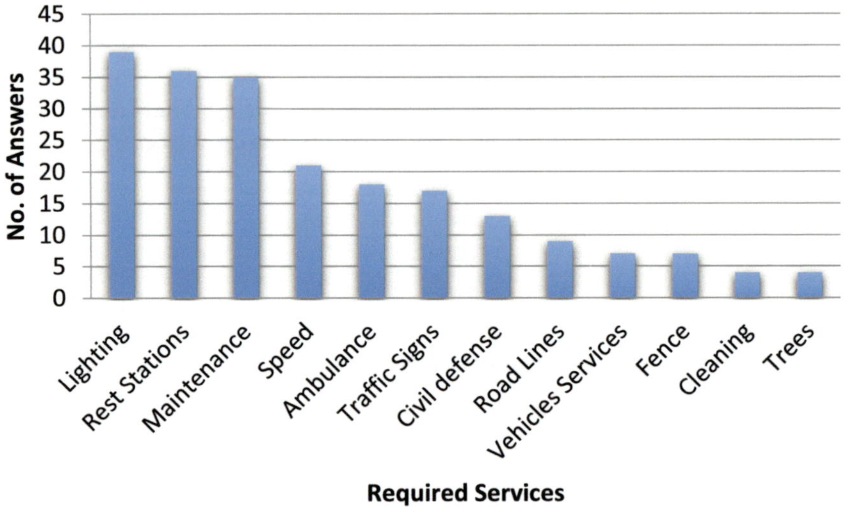


**Figure 2.**
*Services according to the drivers' required priority.*

| Criteria | Evaluate | Sub Criteria | Evaluate |
|---|---|---|---|
| Safety | 9 | Speed | 8 |
| | | Lighting | 4 |
| | | Traffic Signs | 5 |
| | | Pavement Conditions | 6 |
| Road Condition | 8 | Defects | 8 |
| | | Age | 4 |
| | | Maintenance | 7 |
| Services | 7 | Fuel Station | 8 |
| | | Police | 5 |
| | | Ambulance | 4 |
| | | Service Area | 6 |
| Cost | 7 | Travel Time | 8 |
| | | Vehicles Maintenance | 6 |
| | | Vehicles Operation | 5 |
| Geometric Design | 8 | Sight Distance | 7 |
| | | Number of Lane | 7 |
| | | Lane Width | 6 |
| | | Fence | 3 |
| Environment | 6 | Noise | 7 |
| | | Trees | 5 |
| | | Emission | 7 |

**Table 2.**
*Academics evaluation of the importance of primary and secondary standards.*

| Criteria | Current Condition | Toll Condition | Minimum Degree |
|---|---|---|---|
| Safety | 4.7 | 7.7 | 5 |
| Road Condition | 4.3 | 7.5 | 4.5 |
| Services | 3.0 | 6.7 | 3.7 |
| Geometric Design | 5.3 | 7.5 | 4.3 |
| Cost | 5.5 | 7.5 | 4.8 |
| Environment | 4.0 | 7.2 | 4 |

**Table 3.**
*Evaluation scores for the road condition using the main criteria.*

environment by 6 degrees, while road conditions and geometric design were of great importance by 8 degrees, followed by services and cost by 7 degrees. In the same way, the importance and importance of the secondary criteria can be read, as the evaluation was done on a scale from 1 to the least important and 9 to the highest importance.

### 4.3 Expert poll

The experts were asked to evaluate the same six main criteria of the highway in two cases; first represents the current highway condition where the second represents a hypothetical condition when the tolling road system (TRS) is applied **Table 3**. In this table, we note the values of each of the main criteria for the efficiency of the road in its current condition in the first column, while experts expect the values to be as shown in the second column if the TRS is applied, moreover the third column shows the lowest acceptable value for each criterion. It is clear from the table that the efficiency of the road in its current condition is low and requires rehabilitation work at all levels, notwithstanding that the application of the TRS will raise the efficiency of the road for all six criteria.

## 5. Measuring road quality

The current step involves developing a mathematical model to measure the quality of the highway depending on the values of the main criteria previously predicted by the experts.

The highway linking Basra - Nasiriyah has been divided into 28 sections, each section consisting of 10 km in length for the purpose of establishing an assessment from 1 to 9 for the six criteria that determine the efficiency of the road. Out of the total 28 surveyed sections of the highway, 25 sections were used to develop the road quality model while the other three sections were randomly excluded for later use in a cross-validation method to indicate the extent to which the inferred equation represents the actual data. The SPSS program, a software specialized in the statistical analyses, was employed to process the collected data and to find out the best representation of the quality equation based on the main criteria as independent variables. **Figure 3** depicts a screen shot for the data in the SPSS interface.

| Quality | Safety | speed | lighting | traffic_sig. | road_condition | Road_Conditions | defects | age | maintenan |
|---|---|---|---|---|---|---|---|---|---|
| 4.00 | 5.00 | 6.00 | 1.00 | 2.00 | 7.00 | 5.00 | 5.00 | 5.00 | 7.00 |
| 6.00 | 6.00 | 6.50 | 1.00 | 2.00 | 8.00 | 7.00 | 7.00 | 6.00 | 8.00 |
| 4.00 | 4.00 | 5.00 | 1.00 | 1.00 | 4.00 | 4.00 | 4.00 | 3.00 | 5.00 |
| 7.00 | 6.00 | 5.00 | 1.00 | 2.00 | 7.00 | 6.00 | 5.00 | 5.00 | 7.00 |
| 6.00 | 6.00 | 5.00 | 6.00 | 2.00 | 5.00 | 7.00 | 5.00 | 4.00 | 5.00 |
| 5.00 | 4.00 | 5.00 | 1.00 | 2.00 | 4.00 | 7.00 | 5.00 | 4.00 | 5.00 |
| 5.00 | 6.00 | 7.00 | 1.00 | 3.00 | 5.00 | 6.00 | 6.00 | 3.00 | 5.00 |
| 6.00 | 7.00 | 7.00 | 1.00 | 3.00 | 7.00 | 6.00 | 7.00 | 5.00 | 7.00 |
| 6.00 | 5.00 | 5.00 | 1.00 | 2.00 | 7.00 | 5.00 | 6.00 | 4.00 | 6.00 |
| 5.00 | 6.00 | 5.00 | 1.00 | 2.00 | 4.00 | 6.00 | 4.00 | 4.00 | 6.00 |
| 4.00 | 4.00 | 5.00 | 1.00 | 1.00 | 4.00 | 3.00 | 3.00 | 3.00 | 4.00 |
| 6.00 | 7.00 | 7.00 | 1.00 | 2.00 | 6.00 | 5.00 | 5.00 | 4.00 | 6.00 |
| 7.00 | 7.00 | 6.50 | 1.00 | 2.00 | 8.00 | 6.00 | 7.00 | 4.00 | 6.00 |
| 6.00 | 7.00 | 6.00 | 1.00 | 2.00 | 6.00 | 6.00 | 7.00 | 5.00 | 7.00 |
| 4.00 | 5.00 | 5.00 | 1.00 | 2.00 | 4.00 | 4.00 | 3.00 | 8.00 | 2.00 |
| 4.50 | 5.00 | 5.00 | 1.00 | 2.00 | 7.00 | 5.00 | 6.00 | 8.00 | 2.00 |
| 5.00 | 6.00 | 5.00 | 1.00 | 2.00 | 5.00 | 5.00 | 7.00 | 8.00 | 3.00 |
| 4.50 | 5.00 | 5.00 | 1.00 | 2.00 | 5.00 | 5.00 | 6.00 | 8.00 | 2.00 |
| 4.50 | 5.00 | 7.00 | 1.00 | 1.00 | 4.00 | 5.00 | 5.00 | 8.00 | 2.00 |
| 5.50 | 5.50 | 7.00 | 1.00 | 2.00 | 8.00 | 7.00 | 7.00 | 8.00 | 7.00 |
| 4.00 | 5.00 | 5.00 | 1.00 | 1.00 | 3.50 | 4.00 | 4.00 | 8.00 | 2.00 |
| 5.00 | 4.50 | 3.00 | 1.00 | 2.50 | 3.50 | 3.00 | 3.00 | 8.00 | 2.00 |
| 6.00 | 5.00 | 6.00 | 1.00 | 2.00 | 5.50 | 5.00 | 5.00 | 8.00 | 4.00 |
| 5.00 | 4.00 | 5.00 | 1.00 | 2.00 | 4.50 | 4.00 | 4.00 | 8.00 | 3.00 |
| 4.00 | 4.00 | 5.00 | 1.00 | 2.00 | 3.00 | 4.00 | 3.00 | 8.00 | 3.00 |

| Services | fuel_ | police | ambulanc | services_ar. | Cost | travel_time | vehicle_m. | vehicle_operati. | Geometric_design | sight_dist | number_ | lane. | fence |
|---|---|---|---|---|---|---|---|---|---|---|---|---|---|
| 3.00 | 1.00 | 5.00 | 1.00 | 3.00 | 5.00 | 7.00 | 5.00 | 5.00 | 5.00 | 7.00 | 7.00 | 7.00 | 5.00 |
| 4.00 | 1.00 | 2.00 | 1.00 | 2.00 | 5.00 | 8.00 | 7.00 | 5.00 | 6.00 | 9.00 | 7.00 | 7.00 | 3.00 |
| 3.00 | 1.00 | 5.00 | 1.00 | 3.00 | 5.00 | 7.00 | 5.00 | 5.00 | 4.00 | 7.00 | 7.00 | 7.00 | 1.00 |
| 5.00 | 7.00 | 5.00 | 1.00 | 2.00 | 7.00 | 8.00 | 7.00 | 5.00 | 8.00 | 9.00 | 7.00 | 7.00 | 9.00 |
| 4.00 | 7.00 | 5.00 | 1.00 | 2.00 | 5.00 | 8.00 | 5.00 | 4.00 | 8.00 | 9.00 | 7.00 | 7.00 | 9.00 |
| 2.00 | 1.00 | 1.00 | 1.00 | 1.00 | 5.00 | 8.00 | 5.00 | 4.00 | 6.00 | 9.00 | 7.00 | 7.00 | 3.00 |
| 2.00 | 1.00 | 1.00 | 1.00 | 1.00 | 5.00 | 7.00 | 5.00 | 4.00 | 6.00 | 9.00 | 7.00 | 7.00 | 3.00 |
| 2.00 | 1.00 | 1.00 | 1.00 | 1.00 | 6.00 | 8.00 | 5.00 | 4.00 | 6.00 | 8.00 | 7.00 | 7.00 | 3.00 |
| 4.00 | 1.00 | 3.00 | 1.00 | 2.00 | 6.00 | 7.00 | 7.00 | 5.00 | 8.00 | 8.00 | 7.00 | 7.00 | 7.00 |
| 4.00 | 1.00 | 5.00 | 1.00 | 2.00 | 6.00 | 7.00 | 7.00 | 5.00 | 6.00 | 8.00 | 7.00 | 7.00 | 7.00 |
| 3.00 | 1.00 | 1.00 | 1.00 | 1.00 | 6.00 | 7.00 | 7.00 | 5.00 | 5.00 | 8.00 | 7.00 | 7.00 | 3.00 |
| 4.00 | 5.00 | 2.00 | 1.00 | 2.00 | 6.00 | 7.00 | 7.00 | 5.00 | 8.00 | 9.00 | 7.00 | 7.00 | 7.00 |
| 5.00 | 1.00 | 3.00 | 1.00 | 3.00 | 6.00 | 7.00 | 5.00 | 5.00 | 7.50 | 8.00 | 7.00 | 7.00 | 9.00 |
| 4.00 | 1.00 | 2.00 | 1.00 | 2.00 | 6.00 | 7.00 | 7.00 | 5.00 | 6.50 | 8.00 | 7.00 | 7.00 | 7.00 |
| 3.00 | 1.00 | 5.00 | 1.00 | 3.00 | 5.00 | 7.00 | 5.00 | 4.00 | 5.00 | 9.00 | 7.00 | 7.00 | 1.00 |
| 3.00 | 1.00 | 5.00 | 1.00 | 3.00 | 5.50 | 7.00 | 5.00 | 4.00 | 6.00 | 9.00 | 7.00 | 7.00 | 3.00 |
| 4.00 | 1.00 | 5.00 | 1.00 | 3.00 | 6.00 | 6.00 | 6.00 | 6.00 | 6.50 | 9.00 | 7.00 | 7.00 | 1.00 |
| 2.50 | 1.00 | 1.00 | 1.00 | 1.00 | 5.00 | 6.00 | 6.00 | 6.00 | 6.00 | 7.00 | 7.00 | 7.00 | 1.00 |
| 2.50 | 1.00 | 1.00 | 1.00 | 1.00 | 4.50 | 7.00 | 5.00 | 4.00 | 6.50 | 9.00 | 7.00 | 7.00 | 1.00 |
| 4.00 | 1.00 | 5.00 | 1.00 | 2.00 | 6.00 | 7.00 | 8.00 | 8.00 | 7.00 | 9.00 | 7.00 | 7.00 | 3.00 |
| 3.00 | 1.00 | 5.00 | 1.00 | 4.00 | 5.00 | 7.00 | 4.00 | 5.00 | 6.00 | 9.00 | 8.00 | 7.00 | 1.00 |
| 4.00 | 1.00 | 5.00 | 1.00 | 2.00 | 4.00 | 3.00 | 3.00 | 4.00 | 7.00 | 9.00 | 8.00 | 7.00 | 1.00 |
| 4.00 | 1.00 | 2.00 | 1.00 | 2.00 | 6.00 | 6.00 | 6.00 | 6.00 | 7.00 | 9.00 | 8.00 | 7.00 | 1.00 |
| 3.00 | 1.00 | 1.00 | 1.00 | 1.00 | 5.00 | 5.00 | 5.00 | 5.00 | 5.00 | 6.00 | 8.00 | 7.00 | 1.00 |
| 3.00 | 1.00 | 1.00 | 1.00 | 1.00 | 4.00 | 4.00 | 3.00 | 4.00 | 5.00 | 6.00 | 8.00 | 7.00 | 1.00 |

**Figure 3.**
*Data sheet of SPSS program.*

## 5.1 Highway quality model development

After collecting and analyzing the on-site data with the questionnaires that were conducted, there was a linear relationship between the quality of the road and the main criteria, as shown below

$$Q = -0.688 + 0.159S + 0.198Rc + 0.246Sr + 0.185C + 0.270Gd + 0.089E \quad (1)$$

Where:
Q = Quality of the road.
S = Safety.
Rc = Road condition.
Sr. = Services.
C = Cost.
Gd = Geometric design.
E = Environment.

The empirical data of the excluded three expressway sections are utilized for verifying the regression model's output results and for testing its accuracy as shown in **Table 4**.

For the purpose of applying the expert weights to each of the criteria and plugging them into the Eq. (1), the equation becomes:

$$\mathrm{Qc} = -0.688 + (0.159{*}9{*}\mathrm{S} + 0.198{*}9{*}\mathrm{Rc} + 0.246{*}7{*}\mathrm{Sr} + 0.185{*}7{*}\mathrm{C} + 0.270{*}8{*}\mathrm{Gd} + 0.089{*}6{*}\mathrm{E})/9$$

$$Qc = -0.688 + 0.159S + 0.176Rc + 0.191Sr + 0.144C + 0.24Gd + 0.06E \quad (2)$$

Where: Qc is Corrected Quality.

When applying this equation to the values in **Table 2**, it is possible to use the data in the third column that represent the least acceptable values for the main criteria of road efficiency to find the least acceptable amount of road efficiency Qc as bellow:

| Q1 Measured | Q2 Calculated from Eq. (4–1) | Q1-Q2 | % Error \|Q1-Q2\|*100/Q1 |
|---|---|---|---|
| 7.00 | 6.69 | 0.31 | 4.4 |
| 3.50 | 4.36 | −0.86 | 24.5 |
| 6.00 | 5.54 | 0.46 | 7.6 |

**Table 4.**
*Verification results for the developed equation.*

$$Qc = -0.688 + 0.159^{*}5 + 0.176^{*}4.5 + 0.191^{*}3.7 + 0.144^{*}4.8 + 0.24^{*}4.3 + 0.06^{*}4$$
$$Qc = 3.5 \tag{3}$$

Moreover, when applying the values in the first column of **Table 2**, the efficiency of the road can be found at the current conditions as bellow:

$$Qc = -0.688 + 0.159^{*}4.7 + 0.176^{*}4.3 + 0.191^{*}3 + 0.144^{*}5.5 + 0.24^{*}5.3 + 0.06^{*}4$$
$$Qc = 3.7 \tag{4}$$

## 6. Compare alternatives using AHP

A comparison with the present road management system is undertaken in order to suggest the deployment of the Toll Road System TRS. That is, the AHP software is used to apply the values of the primary and secondary criteria for both cases that were collected by specialist academics. The program can address the best situation and yields the best alternative **Figure 4**.

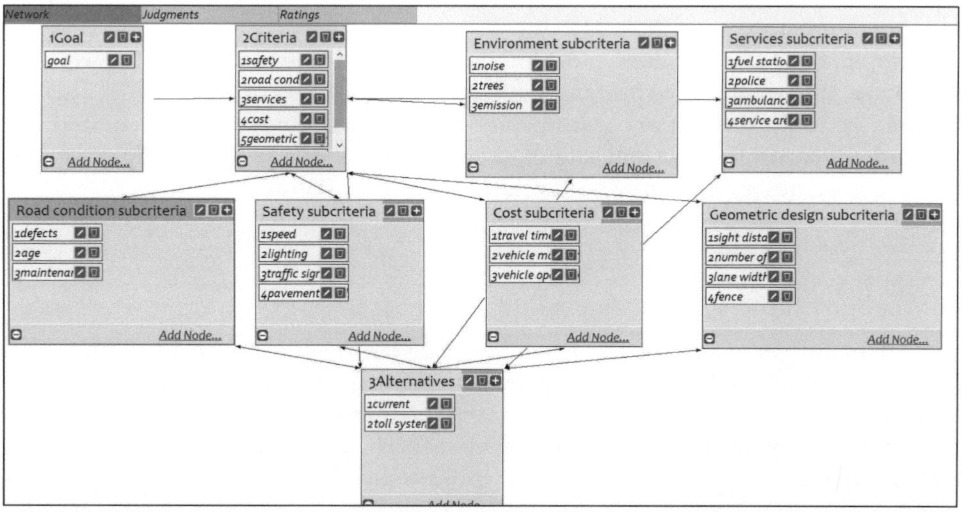


**Figure 4.**
*AHP application interface.*

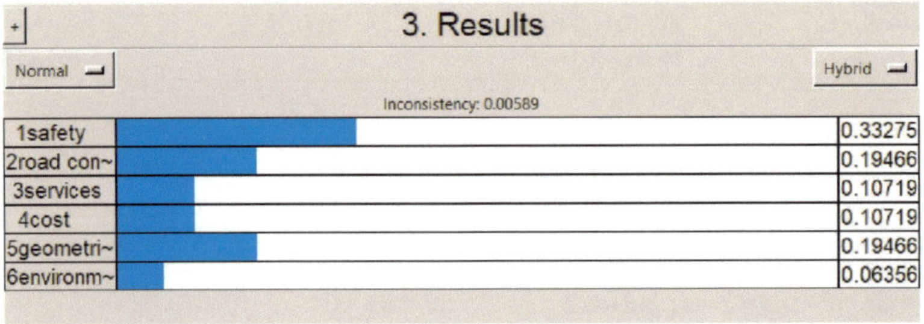


**Figure 5.**
*Numerical values of Main criteria.*

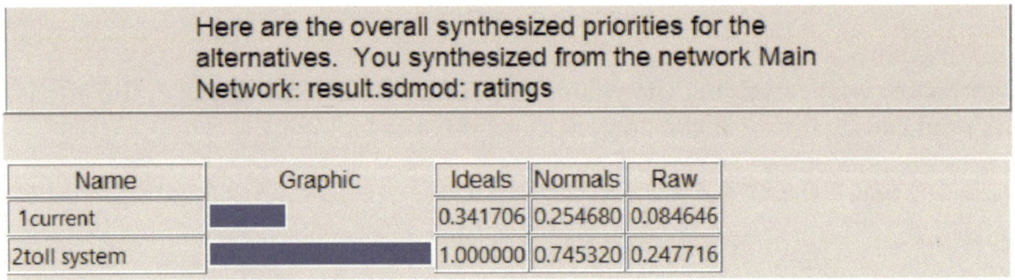


**Figure 6.**
*Synthesis priorities for the alternatives.*

The alternatives are grouped together in a single cluster, here we have two alternatives for highway management, one of them represent the current management system and other for the toll road system TRS.

**Figure 5** shows the numerical values of each of these main criteria and their impact on the quality of the road. It is noted that safety take the largest importance with value of 33.2%, while the road conditions and the geometric design are equal at 19.4%, as well as the services with the cost of 10.7% while the environmental impact is less importance with 6.3%. The amount of inconsistency should be less than 0.1, which is achieved here by a comfortable amount.

After entering all the values and conducting the treatments to find the best alternative, the result was to prefer the toll road system TRS at a rate of 75%. This is an expected and logical result as has how this system could provide all the required standards for a comfortable and safe road operation. As can be seen in **Figure 6**, the differential values are shown in the middle (Normal) column.

## 7. Conclusions

Iraq's roadways management system has to be thoroughly examined and modified. There is currently no defined management agenda in place for gathering data and developing strategies for highway monitoring, maintenance, and service. In Iraq, the risk of accidents is relatively high; according to WHO estimates from 2002, the fatality rate is between 19 and 28 per 100,000 people. This is in addition to the absence of safety measures such as speed limits and other restrictions that apply to road users. The results of the questionnaire survey clearly demonstrated the drivers' unhappiness with the expressway's existing operating performance and quality. As a

*DOI: http://dx.doi.org/10.5772/intechopen.105852*

result, they declare their willingness to pay fees if the expressway's condition is improved. The model designed for judging expressway quality based on the primary and secondary criteria was based on the questionnaire responses. (Eq. 1). The developed road quality equation was modified according to the relative impact of each criterion on the road quality (Eq. 2).

On a 9-level scale, applying this equation to the highway in its current state (Eq. 4) yielded a 3.7 quality standard. The figure is quite close to the minimum of 3.5 (Eq. 3) that experts have set as the limit beyond which maintenance must be performed. According to traffic volume records, the expressway typically traffic more than 10,000 cars every day, half of which are trucks transporting merchandise from Basra port to other Iraqi provinces. A Tolling Road System (TRS) has been proposed for the goal of improving highway management and societal satisfaction, as well as the existence of a legal foundation. The AHP program was used to compare alternatives and, as a result, to pick the best ones by evaluating their application to primary and secondary standards. The results showed that the alternative of implementing the TRS was supported by 75%.

## 8. Recommendations

It's critical to have a permanent system for collecting data and performing surveys along the roadway on a regular basis. This would guarantee that data is available for the implementation of successful maintenance plans. When the necessary data is available, it is useful to expand the usage of the generated road quality model by applying it to additional sections. More case studies are needed to determine the efficacy and practicality of imposing taxes on road users, particularly from a social and economic standpoint.

## Author details

Abdul Ridha Mohammed Afrawee[1]*, Haider Habeeb Aodah[2]
and Hussein Ali Mohammed[1]

1 Civil Department, Kerbala University, College of Engineering, Iraq

2 Civil Department, ThiQar University, College of Engineering, Iraq

*Address all correspondence to: engredha@gmail.com

## References

[1] Dar-Al-Handasah. Iraq Transport Sector Master Plan. Diagnostics Report. Baghdad, Iraq: Ministry of Construction and Housing MOCH; 2014

[2] Choi N, Jung K. Measuring efficiency and effectiveness of highway management in sustainability. Sustainability. 2017;**9**(8):1347

[3] Wheeler AP, Angermeier PL, Rosenberger AE. Impacts of new highways and subsequent landscape urbanization on stream habitat and biota. Reviews in fisheries Science. 2005; **13**(3):141-164

[4] He C, Wang B. Research progress and development trends of highway tunnels in China. Journal of Modern Transportation. 2013;**21**(4):209-223

[5] Mirshahi M, Obenberger J, Fuhs CA, Howard CE, Krammes RA, Kuhn BT, et al. Active Traffic Management: The Next Step in Congestion Management. Federal Highway Administration: United States; 2007

[6] Šelih J, Kne A, Srdić A, Žura M. Multiple-criteria decision support system in highway infrastructure management. Transport. 2008;**23**(4): 299-305

[7] Zhang X, Mao X, AbouRizk SM. Developing a knowledge management system for improved value engineering practices in the construction industry. Automation in Construction. 2009; **18**(6):777-789

[8] Rogers M, Enright B. Highway Engineering. New York: John Wiley & Sons; 2016

[9] Cruz CO, Sarmento JM. The price of project finance loans for highways. Research in Transportation Economics. 2018;**70**:161-172

[10] Sanchez R. Dynamic Pricing on the Road: How Managed Tolls are Increasing Efficiency and Innovation. 2015

[11] Jia YH, Tian ZZ. Toll rates and related factors on China's highways. Transportation Research Record. 2002; **1812**(1):22-26

[12] Serafino N, Tarnoff C, Nanto DK. US occupation assistance: Iraq, Germany and Japan compared. Library of Congress Washington DC Congressional Research Service. March 2006

[13] Al-Neami MA, Al-Rubaee RH, Kareem ZJ. Evaluation of pavement condition index for roads of Al-Kut City. International Journal of Current Engineering and Technology. 2017;7(4): 1461-1467

Chapter 8

# A Drainage System for Road Construction on Flat Terrain

*Owuama C. Ozioma*


## Abstract

Structural integrity of flexible road pavements is guaranteed if an effective road drain is provided during construction. On sloppy terrain open concrete drain has sufficient conveyance to get rid of runoff from the road system. The ability decreases as the terrain approaches near flat condition On flat terrain water often ponds on the roadway after a rainfall and conventional road drainage system is hardly effective in discharging the runoff. To address this a trenchless drainage system becomes a suitable option. This involves an engineered open trench backfilled with granular materials. The method is effective and cheaper.


**Keywords:** drainage, road, flat terrain

## 1. Introduction

Road is an indispensible feature of any developing or developed society. It provides a means of communication, within and in-between communities and cities. It can be a single lane road or dual carriage way which aligns or cuts across with one another. During its construction, provisions are made for an effective road drainage system that will sustain the structural integrity of the road pavement. This measure will guarantee the usability and sustainability of the road for an extended period of time. The drain, located on both or either side of the road, provides a means by which runoff from the road surface and adjacent facilities is discharged harmlessly into an engineered or natural outlet.

The effectiveness of the drainage system is a function of the drain invert slope, its size or capacity and the structural disposition of the lining [1]. For a well designed drain the steeper the slope of the invert the more efficient is the discharge up to a maximum critical slope beyond which the flow becomes supercritical and abrasive. However, as the drain invert decreases to a minimum critical slope the discharge becomes subcritical, less efficient and less self cleansing. As this slope approaches a near flat regime, the potential hydraulic head approaches zero. At this stage the drain conveyance becomes paralyzed. Consequently the drainage system will functionally collapse and becomes grossly inefficient in conveying the runoff, thereby resulting to sustained flooding of the area if rainfall persists. In the same vein, the drain may be silted up, vegetation may spring up to cover the surface of the drain, and the stagnant water becomes a breeding ground for mosquitoes and other water borne disease vectors [2]. Apparently malaria infestations and such health related hazards grow exponentially. Some functionally collapsed open drainage systems on a flat terrain in New Owerri district, Imo state Nigeria are shown in **Figures 1–4** below, as an illustration.

However, the discussion in this work will principally be directed to an engineered drainage system - a trenchless drain, that can permit gradual disposal of road wash/runoff on a flat terrain, after a rainfall.

**Figure 1.**
*A waterlogged open concrete drain overgrown with weeds in new Owerri Nigeria.*

**Figure 2.**
*A waterlogged open concrete drain being used as a waste dump site in new Owerri Nigeria.*

**Figure 3.**
*A waterlogged open concrete drain serving as mosquito breeding ground in new Owerri Nigeria.*

**Figure 4.**
*A stagnant pool of water in an open concrete drain in new Owerri Nigeria.*

## 2. Flat terrain

The geomorphology of an area is defined by its landforms. The landform could be made up of hills, valleys, lowlands and/or flood plains. The aerial disposition of these land features manifest as the topography of the area. That means, the topography or, in other words, the terrain of an area can be undulating, hilly, steep, gentle or flat.It is the topography of a place that determines its natural drainage pattern and this may guide the networking of man made or engineered drainage systems. However, man can alter, to some extent, the existing natural landforms to fit into his developmental objectives.

A flat terrain is a land form with slopes not steeper than 2% [2, 3]. This type of terrain can occur within a region of greater relief such as hills and mountains or in an undulating or alluvial plain. It can also occur on top of a table land forming an escarpment.

The main characteristic of a flat terrain is water logging after a rainfall. In this regard surface runoff remains within the surrounding since there is no sufficient hydraulic head, amidst other obstructions, to drive the flow along. In which case natural drainage is restricted and flooding becomes overwhelming as long as there is a continuous rainfall. This is apparent during the rainy season.

Human settlement is often attracted to flat terrain, probably because the land would not be subjected to pronounced erosion of the soils, agricultural practice is sustainable and existential construction activities are easier to manage. Particularly, road construction on a flat terrain requires minimal earth cutting and earth embankment.

## 3. Road construction

Road construction may involve rigid pavement or flexible pavement. Rigid pavement is made up of portland cement concrete base. It could be reinforced with steel bar or could be an ordinary plain concrete slab.Because of its rigidity and high modulus of elasticity, rigid pavement tends to distribute the applied load over a relatively wide area of soil and the major portion of the structural capacity is provided by the slab. Hence the major consideration in the design of rigid pavement is the strength of the concrete. Minor variations in the sub grade and/or base strength have little influence on the structural capacity of the pavement. Provision of drainage assets is therefore not a priority. In flexible pavement the strength is brought about by building up relatively thick layers of sub base, base course and surface coating over a sub grade. These layers are composed of natural aggregates. The manners of load distribution in flexible pavement make it sensitive to moisture variation. Therefore the provision of drainage asset is a priority in the design and construction of flexible pavement.

In flexible pavement design four main features are considered [2]: the pavement structure, pavement materials, cross sectional profile and size of the drain invert.

### 3.1 Pavement structure

A flexible road pavement structure may comprise the subgrade, subbase, base course and asphalt coating in that order. The subgrade is the in-situ or borrowed earth material spread and compacted in line with the specifications, across the width and lenght of the proposed road to form its foundation. The thickness of the sub grade is semi infinite and the elevation relative to the ground level is dependent on the environmental factors. The subbase is placed over the subgrade and serves as a transition layer to the base course which is the main load spreading layer of the pavement. The uppermost layer is the surfacing or asphalt coating that has a water proofing and abrasion resistant properties.

### 3.2 Pavement materials

The subgrade is comprised of compacted sandy clay, clayey sand or treated mixtures of sand and silt. The subbase may be made of unprocessed sandy gravel or gravelly sand while the base course is composed of specified thickness of crushed stone or gravel stabilized with cement, lime or bitumen. Then the surfacing is comprised of bitumen and often referred to as asphalt.

### 3.3 Cross sectional profile

To prevent rainwater retention at the surface of the pavement the cross sectional profile of the road is seen to be cambered towards the drainage inverts on both sides of the road. This allows the road wash to flow towards the side line of the road.

### 3.4 Drain invert

The drain inverts of the road are positioned along the drainage lines on both sides of the road to convey runoff from the road surfaces towards a sump or discharge basin, down slope. The drainage system could be an open channel or pipe drain depending upon environmental considerations. The conveyance of the drain will be designed based upon rainfall and catchment regimes.

## 4. Drainage systems

The conventional drainage systems commonly adopted in road construction include open concrete drain and pipe drain. In the design of such a drain some factors which include hydraulic, structural, environmental, sociological and maintenance attributes are considered [4]. The weight attached to each of these factors is a function of engineering judgment and sustainability requirement.

### 4.1 Open concrete drain

Open drains can be provided on either or both sides of the road. It can be trapezoidal, rectangular or square in shape. The channel can be lined with mass concrete or reinforced concrete. For the drain to be effective the grade must be sufficient to mobilize minimum permissible velocity that will enhance self cleansing and gentle enough not to exceed the maximum permissible velocity that would initiate scouring of the lining. Open drain can easily be maintained in the face of high sediment load or rubbish load in an area. However it is unsightly and may be risky to road users in some cases. But when the invert slope is less than 2% as is the case with flat terrain, the drain can hardly function. In this situation a trenchless drainage system provides a good alternative.

### 4.2 Pipe drain

Pipe drain is often cylindrical or rectangular in shape, aligned and buried in the subsurface. Drainage through pipes simulates flow through conduits. Pipe drain is mainly suitable for high density and high value residential development. It provides a better appearance and safety than open drain but much more expensive to construct and more challenging to maintain. To function effectively pipe drain must mobilize permissible flow velocities within the range of minimum and maximum. However when the drain invert or the slope of the terrain is less than 2% its conveyance becomes grossly inadequate. Then the alternative and effective choice is the trenchless drain system.

## 5. Trenchless drainage system

The trenchless drainage system provides a solution for the evacuation of runoff from road pavement situated on flat terrain. It serves as a suitable alternative to open or pipe drains on such an environmental setting.

### 5.1 Concept

The concept behind trenchless drainage system is based on the application of engineered absorption field and grass cover to facilitate runoff infiltration into the subsurface [2]. This aligns with the concept of 'French drain' in a general term. The components of trenchless system is highlighted as follows.

**Absorption field**: An absorption field is an excavation or a trench, backfilled with relatively permeable material when compared with the native or in-situ soil. The excavation creates a wider surface area for the water to infiltrate and percolate in the underlying soil medium [5].

**Grasses:** The benefits of grasses as an aid to effective drainage of an area had been highlighted [6, 7]. The foliage intercepts raindrops that attempt to fall unto the ground to cause splash erosion. This, in consequence absorbs the terminal energy of the rain drop thereby reducing its erosivity. The foliage also provides an evapotranspiration medium for the system. The stem of the grass in conjunction with the leaves interrupt free flow of the runoff, thereby increasing the flow roughness. This provides suitable opportunity for infiltration to take place. The root system has numerous channels for water infiltration and as well acts as a transition channel for evapo-transpiration. Generally grasses bind soil particles into a matrix and provide numerous infiltration and evapo-transpiration pathways.

A combination of absorption field and grass in an engineered fashion provides a good infiltration sump for road wash. It also restricts migration of composite earth materials, such as silts and fine sand, in the runoff that could block infiltration channels.

### 5.2 Preliminary studies

The design of trenchless drain requires some fundamental parameters that must be studied and considered. These include rainfall intensity of the area receiving attention, the in-situ or native soil permeability/hydraulic conductivity, properties of the backfill material and the design road width [2].

**Rainfall Intensity**. The rainfall intensity of the area can be obtained from rainfall studies of the nearest metrological station. For optimal design the value may be determined from maximum daily rainfall data of 10 years return period. Rainfall intensity (I) can be calculated from the expression,

$$I = Dr./T \tag{1}$$

where Dr. = depth of rainfall in cm, T = duration of rainfall in min.

**In-situ Soil Properties.** The permeability or hydraulic conductivity of the in-situ soil is required and this can be obtained through a number of simple field techniques. One of such techniques is the Auger Hole method [8], performed within a relatively shallow depth. A number of such a test is carried out along the length and breadth of the proposed road right- of- way. The soil permeability, K is estimated from the equation,

$$K = C\Delta H/\Delta T \tag{2}$$

Where C = the geometry factor of the auger hole, ΔH = change in water level in the hole at time interval $\Delta T$.

Ring infiltration method can also be used in estimating K [9]. It uses the principle of falling head permeameter. The evaluation follows the following equation:

$$K = A_1\,L/A_2\,(T_2–T_1)\;\ln\;(H_1/H_2) \tag{3}$$

Where $A_1$ = cross sectional area of the observation tube, $A_2$ = cross sectional area of the soil tube, L = length of the soil tube, $H_1$ = head at the time $T_1$, $H_2$ = head at the time $T_2$.

In addition to the estimation of soil permeability, other basic properties of the in-situ soil would be studied. These include the grain size distribution, the Atterberg limits, specific gravity, specific weight; etc. They can be used for classification of the soil.

**Backfill materials**. The materials that could be used to backfill the dug trench must have passed through a series or cycles of abrasion and then become stabilized after the erosive components of the material may have been dissolved and removed. This ensures that the likely cementing agents and organic matters had been washed away. Sands and gravels fall into this category. They can be sourced from alluvial deposits or processed to obtain the required properties.

The backfill must be relatively pervious compared to the in-situ soil. Its porosity can be obtained from laboratory measurements conducted on some samples of the backfill. Other laboratory studies that are required may include grain size distribution, specific gravity, and specific weight. Research has shown [10], that poor graded materials with relatively high values of diameter at 30% passing and relatively high values of uniformity coefficient are preferred as backfill.

**Road Width.** The width of the roadway is an input parameter in trenchless drainage design. It is a reference catchment area of the drain. The size may be obtained from preliminary design specifications or working drawing for the construction. Consideration would be given to the effective width of the road for drainage purposes. This may include areas outside the limit of asphalt coating. However the actual width to be used in the sizing of the drain will, principally, be dependent on the judgment of the drainage engineer.

## 6. Design of trenchless drainage system

### 6.1 Design philosophy

The design of a trenchless drain assumes that only the road wash flows towards the sides of the road, and flow down the slope is restricted by the relatively flat nature of the terrain. That means in-flow from outside the road margins is not considered in the formulation of the enabling equation. By synthesizing inflow into the drain and outflow from the drain in the process of infiltration into the adjacent soil, the following equations were obtained [2, 11],

The volume of inflow, Q was given by

$$Q = k_1 1\,(W_r + W_t) \tag{4}$$

While the volume of outflow, i.e. the infiltration capacity, F of the drain was given by

$$F = k_2\,3\,W_t\,K \tag{5}$$

For optimal design, Q = F, and considering a square shaped drainage system, the equation becomes

$$k_1 1\,(W_r + W_t) = k_2\,3\,W_t\,K$$

OR,

$$W_t = k_1 1\,W_r/(3\,k_2\,KT^n – k_1 1) \tag{6}$$

Where $W_t$ = theoretical width of the square drain, 1 = rainfall intensity, $W_r$ = design road width, $k_1$, $k_2$ = coefficients representing losses, generally <1, T = time factor, n = standing time of water pool, K = hydraulic conductivity of in-situ soil.

If the rate at which water flows into the drain is assumed to be the same as the rate it infiltrates into the adjacent soil (outflow from the drain), the water standing time, n will be zero i.e. n = 0, and T = 1. Hence Eq. (6) reduces to,

$$W_t = k_1 1\,W_r/(3\,k_2\,K – k_1 1) \tag{7}$$

However, the factors T, $k_1$, $k_2$ are site dependent and can be established empirically. Ideally, where there are no water losses through evaporation or evapotranspiration, $k_1 = k_2 = 1$

$$W_t = 1\,W_r/(3\,K – 1) \tag{8}$$

$W_t$ is apparently the hypothetical width of the drain and is related to the effective design width, $W_d$ by

$$W_t = W_d\,\eta \tag{9}$$

Where η is effective area factor of the backfill and it is the same as its porosity [12]. By substituting Eq. (9) into Eq. (8)

$$W_d = 1\,W_r/(\eta\,(3\,K – 1)) \tag{10}$$

### 6.2 Design considerations

In the design of the trenchless drainage system specific components are considered. They include the dimensions of the drain, backfill materials, grassing and gang way to user housing.

**Dimensioning.** The dimension of the drain is dependent upon a number of variables which include rainfall intensity, size and profile of the road, the porosity of the backfill and the hydraulic conductivity of subgrade soil [2, 10, 11]. Consequently for a square drain configuration,

$$W_d = 1\,W_r/(\eta\,(3\,K{-}1)) \tag{11}$$

Where $W_d$ = design width/depth of the drain, cm, I = rainfall intensity, cm/sec, Wr = width of the road, η = porosity of the backfill, K = permeability/hydraulic conductivity of the in-situ soil.

For a square drain, in cross section, the **width** ($W_d$) is the same as the depth, and the infiltration surface can be taken to be $3W_d$. The infiltration surface of the drain is its perimeter short of the upper width of the drain.

If a rectangular channel is to be considered the same infiltration surface determined by Eq. (10) will be used in its dimensioning. In such a case,

$$2d + w = 3W_d \tag{12}$$

Where d is the depth of the rectangular channel and w is its width.

For an example:

- If the design width of the square drain is calculated to be 1.5 m based on Eq. (10)
- The infiltration surface = 3 x 1.5 m = 4.5 m.
- For a rectangular drain, if the width, w, of the drain is 1.2 m
- Considering the same infiltration surface of 4.5 m
- The depth, d, of the rectangular drain is calculated from, 2d = 4.5–1.2 = 3.3 m OR d = 1.65 m

That means, the dimension of the rectangular drain, in cross section, will be 1.2 m x 1.65 m as against 1.5 m x 1.5 m for a square drain of the same infiltration capacity when fully mobilized.

The **rainfall intensity** to be used in the design shall represent a peak condition of at least 10 years return period. This creates room for an optimal design since rainfall is the subject of control in this scenario. Rainfall data from nearest, asymptotic and reliable meteorological station is appropriate.

The **width of the road surface** (Wr) is measured on site or taken from the construction drawing. If the road is cambered the road wash will flow unto both sides of the road from its centre line, supplying runoff to the drains on both sides. Thus, Wr that could be used in Eq. (10) will be half of the actual design width of the road to be drained.

The value of the **porosity** (η) of the backfill to be applied in the Eq. (10) could be obtained from laboratory analysis of representative samples using standard methods. Typical laboratory results from various soil types obtained in course of research [10, 11] are presented in **Table 1**.

The in-situ soil **permeability**, (K) can be obtained from the field using standardized methods [8, 9]. The measurement can be conducted on a number of locations on the right-of-way of the proposed road project. Measurements can also take place when the subgrade of the road pavement has been prepared. The choice is based on site conditions, construction philosophy and/or judgment of the drainage engineer. The in-situ soil properties could be determined to aid in the soil classification. Results obtained on a similar exercise are shown in **Table 1**.

### 6.3 Backfilling

Backfilling along the length and breadth of the trench is done with suitable materials. Such materials may be sourced from alluvial deposits and treated, if necessary, to satisfy the design requirements. Suitable materials must have been stabilized after long transportation in fluvial regime amidst dissolution and abrasion of original rock. Granite is not recommended. From research finding [10, 11], gravelly sand is most recommended, followed by fine sand and then mixed sand in that order. The results of such evaluations are graphically illustrated in **Figures 5** and **6**. The diameter at 30% passing ($D_{30}$) and Uniformity coefficient (Cu) of a suitable material must be relatively high, **Table 1**.

### 6.4 Grassing

Grass species suitable for use are locally available in abundance and the type to be introduced can be decided by the horticulturist attached to the project. The grass is provided and nourished as a cover over the backfill. The roots of the grasses serve as the fiber reinforcement to the loose backfill and as well provide infiltration channels into the fill. This may prevent clogging of the interstices by oxides and clays suspended in the runoff. In addition, it adds to the esthetic value of the drain.

| **Soil** | **Soil category** | | | |
|---|---|---|---|---|
| **Parameters** | **Mixed sand** | **Fine sand** | **Gravelly sand** | **In situ soil** |
| Water content, % | 13.7 | 18.2 | 9.3 | 32 |
| Void ratio | 0.13 | 0.08 | 0.23 | 0.37 |
| Porosity | 0.12 | 0.07 | 0.19 | 0.27 |
| Bulk wt, kN/m$^3$ | 19.2 | 18.7 | 17.9 | 19.8 |
| Dry unit wt, kN/m$^3$ | 16.9 | 15.8 | 16.4 | 15.0 |
| Sat unit wt, kN/m$^3$ | 20.4 | 19.5 | 19.9 | 20.7 |
| $D_{10}$, mm | 0.3 | 0.3 | 0.6 | 0.26 |
| $D_{30}$, mm | 0.45 | 0.52 | 1.2 | 0.4 |
| $D_{50}$, mm | 0.75 | 0.72 | 2.1 | 0.57 |
| $D_{60}$, mm | 0.82 | 0.8 | 2.6 | 0.73 |
| Cu | 2.43 | 2.67 | 4.3 | 2.81 |
| Cc | 0.82 | 1.13 | 0.9 | 0.84 |
| USC Class | SP (cmg) | SP (cmf) | GP (gcm) | Clayey sand |

**Table 1.**
*Properties of various soil types obtained from laboratory studies.*

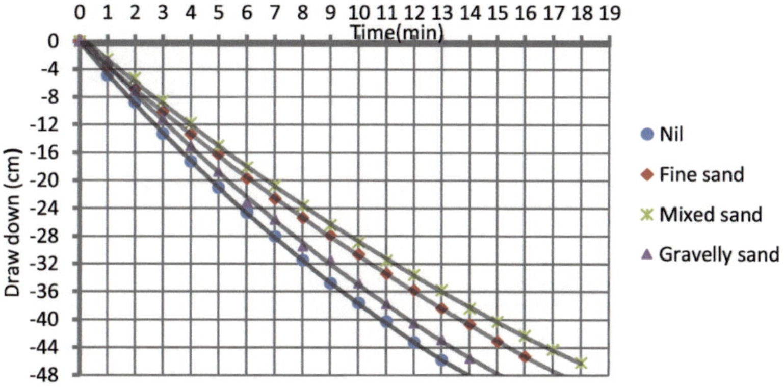


**Figure 5.**
*Drawdown versus time curve with varying backfill materials.*

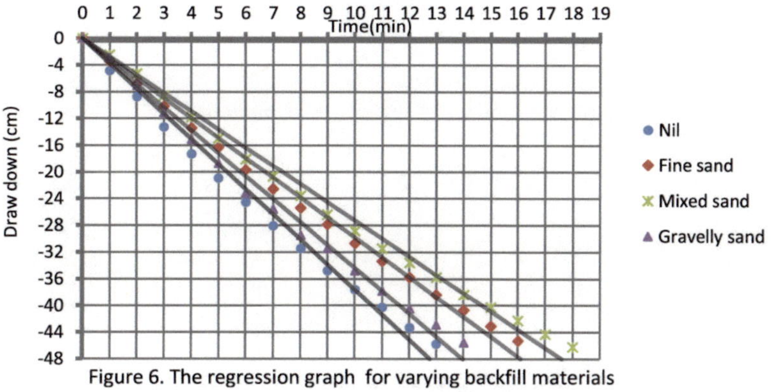


**Figure 6.**
*The regression graph for varying backfill materials.*

### 6.5 Gang way

Situations may arise where the trenchless drain must be bridged. This is mainly in residential areas in which case crossings to serve as entrances to housing units must be provided. This may require placement of precast concrete slabs over the drain and will be limited specifically on the required position.

### 6.6 Comprehensive outline

In special consideration of all the input variables, a hypothetical design of trenchless drainage system on a road section is illustrated in **Figure 7** while the plan of the sketch drawing is shown in **Figure 8**. The alignments and relative locations of the features are defined in the drawing.

## 7. Construction method

The construction of trenchless drain in a road project simply involves trenching, backfilling and grassing.

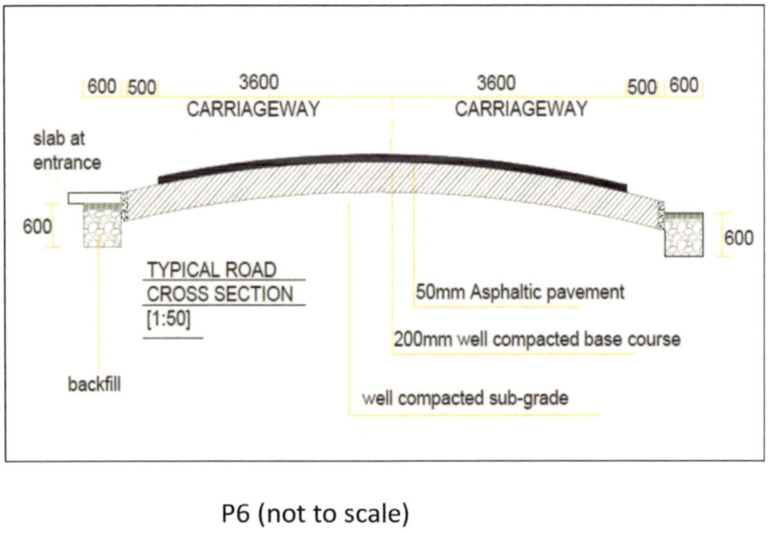


**Figure 7.**
*Hypothetical design section of a trenchless drainage system on a road profile.*

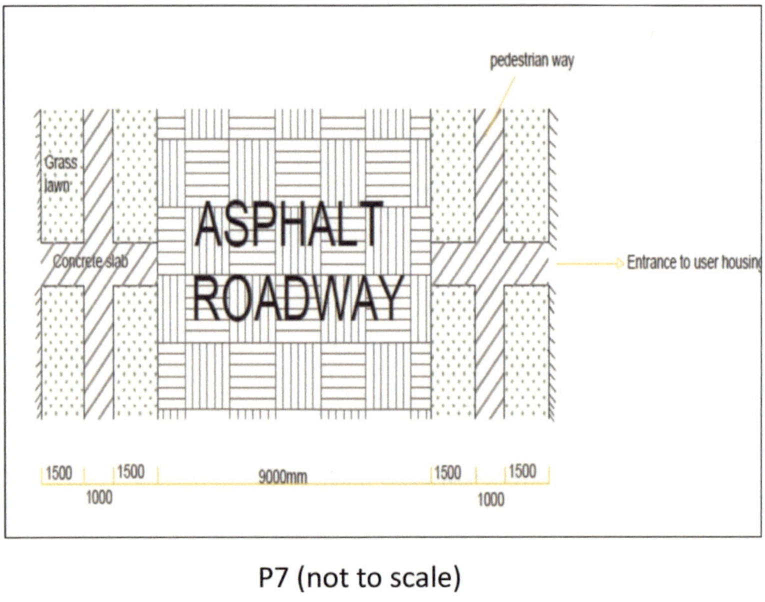


**Figure 8.**
*Plan view of a hypothetical trenchless drainage system in residential district.*

## 7.1 Trenching

This requires the digging of trench along the length of the road on either side or both sides of it in accordance with the specification. Either of motorized excavator or manual labour can be used in the exercise depending upon availability, cost and size of the project. The excavated materials can be disposed as waste or reused depending upon the properties of the soil.

### 7.2 Backfilling

Alluvial sands that can serve as backfill are often locally available. The material could be brought to site by trucks and processed to satisfy the design specifications. It is also possible to lift material that was naturally sorted to meet the requirement. The material can then be dumped and spread in full dimension of the trench to its design level. Densification of the fill is not encouraged. The backfill serves as a transmission medium to the in-situ soil.

### 7.3 Grassing

The last stage in the construction exercise is the introduction of grasses. Specific species of grasses are obtained and planted over the backfill in a specified manner. To sustain its survivability recommended nurturing procedure must be followed. The grass draws in the water into the backfill.

## 8. Comparative advantages

The benefits of the Trenchless drain over conventional methods in road networking on flat terrain can be assessed using the following parameters.

### 8.1 Construction cost

Open concrete drain is constructed of a mixture of aggregates and cement at a design ratio, and sometimes reinforced with iron bars. For a similarly sized trenchless drain the only construction material is sand or gravel put in place with no mandatory skill. A comparative cost analysis for the various dimensions and composition of drains using local market prices was carried out and the results are illustrated in **Figure 9**. It is observed, for example, that for a given dimension, a reinforced concrete drain is about four times (4x) more costly than a trenchless drain constructed of sand aggregate, outside labour cost. Open concrete drain requires properly designed culvert

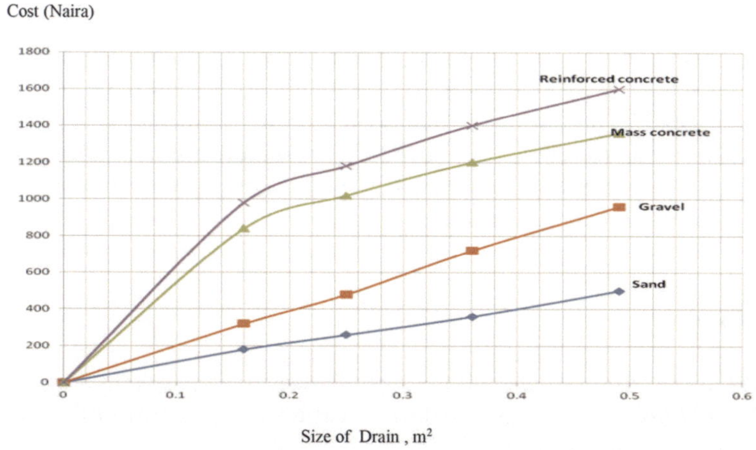


**Figure 9.**
*Cost comparison for various sizes of drains and component materials.*

at each crossing as against slab, if a trenchless drain is constructed. A culvert of known dimension is 10 times more costly than slab of the same size, [11].

### 8.2 Maintenance cost

A concrete drain requires regular cleansing for free flow of water to be sustained. But because of high cost of labour and equipment for the evacuation of the attendant waste, the maintenance could be done ones in a year and preferably at the onset of the rainy season. Comparatively trenchless drain requires grass cutting when overgrown to maintain its environmental friendliness. The maintenance is easily done at a desired time and at a relatively very cheap operational cost.

### 8.3 Aesthetics

Open concrete drain presents an unsightly view especially when it is not properly maintained. Sometimes it serves as an easy waste dump site for uninformed and carefree inhabitants. Comparatively, the greenish outlook of a trenchless drain merges with nature to beautify the environment and provides a healthy and sustainable scenario.

### 8.4 Flooding

Concrete drain is easily submerged in course of rainfall due to obstructions and low grade terrain. The water may remain stagnant for a good length of time, providing a good breeding ground for mosquitoes. This often results to exponential increase in malaria cases. Trenchless drain does not retain flood water for a period of over 1 hour after a torrential rainfall [12]. It therefore provides minimal opportunity for mosquito breeding. On heavily flooded roadway it is often difficult for road users to identify the limit of an open concrete drain and such persons may run the risk of falling into the drain. Trenchless drain provides, apparently, a continuous surface cover with no perceived danger to road users.

### 8.5 Erosion

Where an open concrete or pipe drain is functional its discharge can cause erosion at the outlet of the drain or its culvert, especially where the drain or culvert is not properly terminated. Over 240 well developed gully erosion sites in South East of Nigeria were observed to have been caused by wrong termination of drains and culverts [13]. The trenchless drain can hardly yield significant surface runoff that can initiate any form of erosion down slope.

## 9. Limitations

- *The trenchless drain is suitable for areas flatter than 2% grade. For slopes greater than 2% and less than 7% special design features may be introduced. It is not applicable for slopes greater than 7%.
- The system is not suitable in areas with very low permeability soils such as fat clay

- Where groundwater level is close to the surface the method is ineffective. If hard pan or rock outcrop is shallower than 1 m, the effectiveness of the drain is in doubt.

- The concept does not accommodate deluge of discharge from built – up environment with impermeable land cover. Such land covers may include interlocking stone, asphalt or concrete slab. For instance, a rainfall intensity of 100 mm/hr. on a plot of built up area (33 m by 20 m) can generate 66 cubic meters of water per hour. To handle flow of this magnitude special design consideration may be involved.

## 10. Further research

To further this work the following areas may be of interest for investigation

- Establishment of empirical factors T (time factor), $k_1$ (inflow loss coefficient), $k_2$ (infiltration loss coefficient) for varying climatic conditions.

- Evaluation of cases where the terrain slopes beyond 2% and up to 7%,

- Adaptation of the concept to accommodate slopes steeper than 7%,

- Evaluation of its applicability in low permeability soils.

- Determination of effective absorption surface in the trenchless drain.

## 11. Conclusion

Construction of road network in an urban or semi – urban settlement requires a good drainage system that can convey rainfall surface runoff from impermeable surfaces created by road surface and built up areas. The efficiency of such a drain depends upon the road/invert grade, construction standard and ethics adopted, and maintenance culture.

On flat areas open concrete and pipe drains are hardly efficient in the disposal of rainfall runoff, and are unsustainable especially in a developing economy. A sustainable drainage system in such an environment is the trenchless drain with inbuilt drainage facilities comprising water absorption unit and grass cover. Field and laboratory observations have shown that the system is low cost both in construction and maintenance. It disposes accumulated surface water soon after a rainfall and reduces the incidence of erosion downstream since potential concentrated runoff is eliminated. It also introduces aesthetics values to the environment, and an antidote to mosquito breeding which consequently reduces the incidence of malaria. It is not suitable in sloping area and in districts with impermeable soil layer just below.

Built up environments or housing units could be provided with articulate arrangement of trenchless drain, absorption units and green cells within the premises to minimize the volume of water discharge onto the roadway. This concept will reduce the quantum of water concentrating to form a deluge that may ultimately result to urban flooding, fluvial flooding and downstream erosion. The articulate arrangement of trenchless drain and comprehensive absorption units is an area of research and development interest.

## Acknowledgements

The author wishes to acknowledge the assistance of TETFUND, a Federal Government of Nigeria sponsored funding agency, for providing the enabling grants for the research that buttressed this discussion. The effective participation of all academic and non- academic staff of the Institute of Erosion studies in the Federal University of Technology Owerri Nigeria, at the various segments of the research, is highly appreciated. My secretary, Ms. Rita Ugwu, who meticulously prepared this manuscript is highly esteemed.

## Author details

Owuama C. Ozioma
Department of Environmental Management, Federal University of Technology, Owerri, Nigeria

*Address all correspondence to: chukwunonye.owuama@futo.edu.ng; owuamco@yahoo.com

## References

[1] Van Dort JA, Bos MG. Main drainage systems, Chap IV. In: Design and Management of drainage systems. Wageningen: ILRI press; 1980. pp. 161-171

[2] Owuama CO. Conceptual design of a trenchless drain in a flat terrain. Journal of Environmental Science and Engineering A. 2012;**1**(12) David Publishing USA:1301-1307

[3] Raadama S, Schulze FE. Surface field-Drainage System, design and management system, Chapter IV. In: Drainage Principles and Applications. IILRI; 1980. pp. 69-89

[4] MSMA. Urban storm water management. Vol. Volume I and II. Kuala Lumper: Department of Irrigation and Drainage Malaysia (Reproduced by IES); 2015

[5] Design Manual for Roads and Bridges. Volume 4, Drainage Part 5. Determination. UK Highway Agency. Available from: http://www.standardsforhighways.co.uk/dmrb/vol4/section2/ha4001.pdf [Accessed: October 25, 2012]

[6] Coppin NJ, Richards IG. The use of vegetation in Civil Engineering. London: Butterworths; 1990. p. 365

[7] James AH, Birch P, Palmer J. Land Restoration and Reclamation. Principles and Practice. Available from: www.books.google.com.ng [Accessed: October 25, 2012]

[8] Oswaldo PLS, Stephanie SA, Gerlange SDS, Aline BS, Aivara ISP. Determination of hydraulic conductivity by auger hole method. Biofix Scientific Journal. 2018;**3**(1):91-95

[9] Falling Head Test. Available from: www.trenchlesspedia.com, 2020

[10] Owuama CO, Owuama KC. A drainage system for Road Construction on flat terrain in New Owerri Nigeria. In: Proc. 2nd International Conference, EGRWSE – 2019, University of Illinois, Chicago. IL, USA: Sprinkler publications; 2020

[11] Owuama CO, Uja E, Kingsley CO. Sustainable Drainage System for Road Networking. International Journal of Innovation, Management and Technology. 2014;**5**(2) IACSIT Press: 83-86

[12] Institute of Erosion Studies. Evaluation of Sustainability of Trenchless Drainage System in a Humid Environment. IES Technical Bulletin. Owerri, Nigeria: FUTO Press; 2013. p. 36

[13] Owuama CO. Environmental Consequences of Poor Runoff Management in S.E. Nigeria. Journal of Environmental Systems. 1997;**25**(3) Baywood N. Y:287-302